Praise from the Experts

"I was very excited when I was approached to review *The Global English Style Guide: Writing Clear, Translatable Documentation for a Global Market*. I was even more excited when it arrived and lived up to my hopes. The guide is both comprehensive and succinct, and best of all, is full of practical examples showing text before and after it has been disambiguated. That means there finally is the definitive resource that has been lacking in the field of writing and editing for an international audience."

Wendalyn Nichols
Editor of *Copyediting* newsletter and editorial trainer

"I am amazed by the depth of the analysis and the quality of the examples. I cannot begin to imagine the number of hours required to present such an exhaustive and detailed study.

"Some things that I particularly like include the attention to non-native speakers reading in English, the emphasis on the importance of syntactic cues, and the research presented in the syntactic cues appendix. I greatly appreciate having such an abundance of references identified for me."

Susan Ledford
Master Teacher
Technical Editor

"Backed by solid research and practical industry experience, Kohl's book is a useful, accessible guide with a common sense approach to Global English. I recommend it as a valuable resource to all globalization professionals."

Bev Corwin
Enso Company Ltd.

"This book addresses the growing awareness that technical documents must reach a wider audience than native English speakers: those who read a translated version of the documents and those for whom English is a second language.

"John Kohl's discursive style is informative and instructive, without being labor-intensive or didactic. His flowcharts on revising noun phrases and his discussions on the technicalities of machine translation and the benefits of syntactic cues are presented in an easy-to-understand manner.

"*The Global English Style Guide: Writing Clear, Translatable Documentation for a Global Market* is definitely a "must-have" for anyone who writes for international audiences."

Layla A. Matthew
Technical Editor

The Global English Style Guide

Writing Clear, Translatable Documentation for a Global Market

John R. Kohl

The correct bibliographic citation for this manual is as follows: Kohl, John R. 2008. *The Global English Style Guide: Writing Clear, Translatable Documentation for a Global Market.* Cary, NC: SAS Institute Inc.

The Global English Style Guide: Writing Clear, Translatable Documentation for a Global Market

ISBN 978-1-59994-657-3

SAS Institute Inc., SAS Campus Drive, Cary, North Carolina 27513.

1st printing, March 2008
2nd printing, July 2008
3rd printing, December 2009

Contents

Preface

Is This Book for You?

This book is intended for anyone who uses written English to communicate technical information to a global audience. For example, members of the following professions will find this book especially useful:

- technical writers
- technical editors
- science writers
- medical writers
- proposal writers
- course developers
- training instructors

Even if producing technical information is not your primary job function, the Global English guidelines can help you communicate more effectively with colleagues around the world. By following these guidelines, you can make your e-mail messages and other written communications more comprehensible for colleagues who are non-native speakers of English, whether those colleagues are down the hall or halfway around the world.

This book is also intended to help raise awareness of language-quality issues among managers in the above professions and among Web-site administrators. These individuals should be especially interested in the potential for using language technologies such as controlled-authoring software and machine-translation software to facilitate translation and to improve the quality of translated information. Many companies have found that these technologies produce excellent returns on investment. Those returns are significantly higher and faster when the quality of the English source material is ensured by following the guidelines that are explained in this book.

Localization companies and translators, who have long been among the strongest advocates for language quality, can use this book to help educate their customers. The Global English guidelines help illustrate the fact that the quality and consistency of the source text, not the skill or competence of the translator, are often the biggest factors that affect translation quality.

The Global English guidelines were developed with technical documentation in mind. Nevertheless, most of the guidelines are also appropriate for marketing materials and for other documents in which language must be used more creatively, informally, or idiomatically. If you produce those types of information, many of the guidelines will help you communicate your brand identity clearly and consistently to a global audience.

The Scope of This Book

As its title suggests, this book is a style guide. It is intended to supplement conventional style guides, which don't take translation issues or the needs of non-native speakers into account. It is not a replacement for a technical writing textbook, because it doesn't cover basic principles or guidelines for technical communication.

Instead of attempting to cover every guideline that could conceivably be of interest to anyone who is writing for a global audience, I have focused on the types of issues that I know the most about: sentence-level stylistic issues, terminology, and grammatical constructions that for one reason or another are not suitable for a global audience.

In this book you will find dozens of such guidelines that you won't find in any other source, along with explanations of why each guideline is useful. Often I include specific revision strategies, as well as caveats that will help authors avoid applying guidelines incorrectly or inappropriately.

The amount of explanation that I provide might be more than some readers need. However, in my experience, authors ignore guidelines that are not explained adequately. For example, at many organizations hardly anyone pays attention to the "use active voice" guideline or to the equally vague "avoid long noun phrases" guideline that I have seen so often in other publications. In many technical documents, it simply is not feasible to use only active voice or to avoid *all* long noun phrases. More explanation is required, so I provide such explanations in this book.

In any case, I abhor overgeneralizations and oversimplifications, and I tend to analyze things to death. As you read this book, I hope you will view that personality trait as an asset rather than as a shortcoming!

About the Examples

Because I have worked as a technical writer and editor only in the software industry, most of the example sentences in this book are from software documentation.

If an example sentence has a check mark or a check plus (✓+) next to it, then it conforms to the Global English guidelines. However, please note the following:

- I often did not change passive verbs to active voice. As explained in guideline 3.6, "Limit your use of passive voice," Global English doesn't prohibit the use of passive voice, but recommends using active voice when it is appropriate. A change from passive voice to active voice often causes an unacceptable change in emphasis.
- I did not change terms that are ambiguous in some contexts, but which are unambiguous in the example sentence. For example, I changed *since* to *because* and *once* to *after* or *when* only if there was a potential for misunderstanding.
- I did not follow prescriptivist rules against ending sentences in prepositions, splitting infinitives, and so on, unless I thought that applying one of those rules would improve the example sentence.
- The example sentences are intended to illustrate the Global English guidelines, not to convey technical information accurately. Therefore, I did not hesitate to change example sentences in ways that might make them technically incorrect. For example, I simplified some sentences in order to eliminate distracting problems or unnecessary technical details that were not germane to the discussion. I also changed some of the most obscure software terms to terms that are more likely to be familiar to some readers.

Having said all that, if you have suggestions for improving any of the examples, or if you want to contribute examples of your own, your feedback is welcome. Please contact me via the book's Web site, http://www.globalenglishstyle.com.

Global English on the Web

The Web site for this book, http://www.globalenglishstyle.com, contains additional content that is likely to change or that for various reasons could not be included in the book. Be sure to visit the Web site periodically, and feel free to provide feedback on how the book or the Web site could be improved.

Terminology

In this book, all but the most basic grammar terms are defined or explained in context. The definitions are also included in the glossary. Because the explanations of the guidelines are accompanied by examples, you will be able to understand most of the guidelines without necessarily mastering this terminology.

Terms such as *localization*, which some readers might not be familiar with, are also defined in the glossary.

Acknowledgments

I would like to thank several managers at SAS Institute for recognizing the value of the Global English guidelines and for approving the publication of this book. Those managers include Kathy Council, Gary Meek, Sean Gargan, and Julie Platt in the Publications Division, and Patricia Brown in the SAS Legal Department. Thanks also to Helen Weeks and Pat Moell, co-managers of the Technical Editing Department, for their support and encouragement.

Working with my colleagues at SAS Press, many of whom I have known and worked with on other projects for years, has been much easier than if I had had to work with an outside publisher. Because I was the author instead of the copy editor this time, I gained a better understanding of the innumerable steps that go into producing a book. My acquisitions editor, John West, has done an awesome job of shepherding the project along, attending to myriad details, and patiently adjusting the schedule again and again as I tried to deliver the goods!

I also appreciate the efforts and contributions of many other colleagues at SAS who were involved in the design and production of this book. They include Mary Beth Steinbach, Candy Farrell, Patrice Cherry, Jennifer Dilley, Ashley Campbell, Lydell Jackson, and John Fernez.

My copy editor, Kathy Underwood, went far beyond her usual role, giving me invaluable input that helped shape the content and layout of the book. As she knows, I think of her fondly as "the font of all wisdom." As an editor and as a friend, she is absolutely top-notch!

When my principal reviewers, Mike Dillinger and Jeff Allen, agreed to review this book, I almost could not believe my good fortune. I cannot imagine anyone whose opinions and input I'd have valued more highly than theirs, and I greatly appreciate the time that they devoted to this book. Thanks also to Johann Roturier, Sabine Lehmann, and Helen Weeks for giving me valuable input on parts of the book.

On many occasions I turned to Ronan Martin, a SAS localization coordinator in Copenhagen, for input on the extent to which certain issues posed problems for translators. He often polled the SAS translators and gave me information that was essential for completing parts of this book. Thanks, Ronan, and thanks also for your contributions toward the terminology management initiative at SAS.

Many individuals have encouraged me to write this book. Four of my strongest advocates have been Amelia Rodriguez, Bev Corwin, Mike Dillinger, and Leif Sonstenes. Sue Kocher, SAS terminologist, has been my greatest ally (and one of my best friends) at SAS. Along with Elly Sato and Manfred Kiefer, she has played a big role in the effort to garner support for the Global English guidelines at SAS.

I also want to thank all of my fellow editors and technical writers at SAS who have supported and encouraged me over the years. I feel blessed to work with such a great bunch of people!

Finally, thanks to the librarians at the SAS Library for supporting my research by obtaining the numerous articles and books that I've requested. Of all the great benefits and services that SAS provides to employees, the SAS Library is near the top of my list!

Chapter 1

Introduction to Global English

What Is Global English?

In this book, Global English refers to written English that an author has optimized for a global audience by following guidelines that go beyond what is found in conventional style guides.

The Global English guidelines focus on the following goals:

- eliminating ambiguities that impede translation
- eliminating uncommon non-technical terms and unusual grammatical constructions that non-native speakers (even those who are quite fluent in English) are not likely to be familiar with
- making English sentence structure more explicit and therefore easier for non-native speakers (as well as native speakers) to analyze and comprehend
- eliminating unnecessary inconsistencies

Because Global English doesn't impose severe restrictions on the grammatical constructions or terminology that are permitted, it is suitable for all types of technical documentation.

Why Global English?

The Global English guidelines enable writers and editors to take the clarity and consistency of technical documents to a higher level, leading to faster, clearer, and more accurate translations.

Global English also makes technical documents that are not slated for translation more readable for non-native speakers who are reasonably proficient in English.[1] After all, many documents are never translated, and in today's world it is unusual for the audience of any technical document to consist solely of native speakers of English. Whether your audience consists primarily of scientists, engineers, software developers, machine operators, or unskilled workers, it probably includes a sizable number of non-native speakers.

Finally, Global English makes documents clearer and more readable for native speakers, too. Because native speakers of English still constitute the majority of the audience for many technical documents, that benefit should not be overlooked.

Depending on the type and subject matter of your documentation, Global English can provide the following additional benefits:

- Injuries, losses, and costly legal liabilities that can be caused by unclear documentation and by incorrect translations are avoided.
- Clearer, more-consistent documentation reduces calls to technical support.
- Consistent terminology facilitates the task of indexing and makes indexes more reliable.
- In online documentation or Help, users are better able to find the information they need because you have eliminated unnecessary synonyms and variant spellings.
- Translation quality is less of a concern because you have eliminated ambiguities and unnecessary complexities that can lead to mistranslations.

Benefits of Global English for Professional Writers and Editors

Anyone who produces technical documentation for a global audience should follow the Global English guidelines. However, many professional writers and editors have recognized two benefits of Global English that are less relevant to authors whose main responsibilities are in other areas.

[1] If your audience has limited proficiency in English, then consider using a form of controlled English in addition to following the Global English guidelines. See "What is the relationship between Global English and controlled English?" on page 14.

First, the ability to make documents more suitable for a global audience is a specialized and marketable skill. In a posting to a Society for Technical Communication mailing list, technical writer Richard Graefe made the following observation:

> *With the increase in localization[2] of documentation and of user interfaces, being able to sell yourself as a person with "pre-localization" editorial skills is a plus. To be able to do that type of editing well, you need . . . to be able to recognize English structures and expressions that will not translate well, that may be ambiguous to a translator, or that may require a translator to do rewriting in addition to translating.*

The skills that Graefe described are exactly what this book helps you develop. In addition to giving job seekers a competitive edge, those skills could conceivably make professional writers and editors less vulnerable to layoffs and outsourcing.

Second, editors often find that the Global English guidelines articulate issues that they could not have explained themselves. The same is true for writers who are working with technical information that was provided by subject-matter experts. By referring authors to the explanations in this book, you can often persuade them to make the necessary changes with less resistance or discussion.

The Cardinal Rule of Global English

As noted above, native speakers of English probably constitute a significant portion of the audience for much of your technical documentation. Therefore, be sure to follow the cardinal rule of Global English even while you are taking into account the needs of non-native speakers and translators:

> **The Cardinal Rule of Global English**
>
> Don't make any change that will sound unnatural to native speakers of English.

[2] *localization*: the process of adapting products or services for a particular geographic region or market. Translation is a large part of the localization process.

At the same time, consider the following corollary to the cardinal rule:

> **Corollary**
> There is almost always a natural-sounding alternative if you are creative enough (and if you have enough time) to find it!

In other words, if following one of the Global English guidelines would cause a sentence to sound stilted or unnatural, then either find a different way to improve the sentence, or leave the sentence alone.

Consulting Colleagues

If you are a non-native speaker of English, your instincts about what sounds natural in English and what doesn't might not always be reliable. If you are not sure whether you are following the cardinal rule successfully, consult a native speaker whose judgment you trust.

Native speakers also benefit from consulting other native speakers on occasion. A colleague might quickly find one of those "natural-sounding alternatives" that eluded you.

Global English and Language Technologies

Often, people who are interested in Global English are also interested in technologies that make global communication more efficient. Three language technologies that are mentioned frequently throughout this book are machine-translation software, translation memory, and controlled-authoring software. The following sections provide an overview of these technologies and of how they relate to Global English.

Machine-Translation Software

Machine-translation (MT) software is software that translates sentences from one language (such as English) into one or more other languages (such as French or Japanese).

This book doesn't include guidelines that are useful *only* for improving the output of machine-translation software. However, many of the Global English guidelines that make documents more suitable for translation by human translators also make documents more suitable for machine translation. A sentence that is unclear, ambiguous, or otherwise problematic for human translators is often translated incorrectly by machine-translation software.

On the other hand, it is important to note that relatively little research has been done on the effect of specific style guidelines and terminology guidelines on machine-translation output. A guideline that improves the translated output for one MT system in one language might have a negligible effect (or, rarely, even a detrimental effect) for a different MT system or language.

Before implementing machine-translation software, ask the software vendor to help you identify and prioritize the Global English guidelines that would make your documentation most suitable for that particular software. Also consult sources such as Bernth and Gdaniec (2000), Roturier (2006), and O'Brien and Roturier (2007) to gain a better understanding of how to evaluate the effect of specific guidelines on MT output.

Case Study

A small pilot project conducted at SAS Institute indicated that the Global English guidelines have a significant effect on the quality of machine-translation output. This study did not examine the effect of specific guidelines. Instead, the project coordinators took a subjective look at the effect of following all of the Global English guidelines that were in the *SAS Style Guide for User Documentation* as of December 2004.

In this project, SYSTRAN translation software was used to translate the documentation for one software product from English to French. The process consisted of these steps:

1. Technical terms were pre-translated and added to the SYSTRAN dictionary.
2. A small part of the document was translated without being edited first.
3. The same part of the document was edited according to the Global English guidelines.
4. That part of the document was translated again.
5. The translations of 22 sentences were evaluated by professional translators.

As Table 1.1 shows, the translations of the Global English version of the document were significantly better than the translations of the unedited version. The percentage of sentences that were rated as either Excellent or Good increased from 27% to 68%. The percentage of sentences that were rated as either Medium or Poor decreased from 73% to 32%.

Table 1.1 Evaluations of Translations Produced by SYSTRAN, English-French

Translations of **Unedited** Sentences			Translations of **Edited** Sentences		
Rating	Number of Sentences	Percentage	Rating	Number of Sentences	Percentage
Excellent	0	0.00	Excellent	6	27.27
Good	6	27.27	Good	9	40.91
Medium	13	59.09	Medium	7	31.82
Poor	3	13.64	Poor	0	0.00

The sample size was admittedly very small. However, the results are consistent with results reported by Roturier (2006) and with what common sense tells anyone who has worked with computers: the quality of the output depends largely on the quality of the input.

More and more large companies are using machine translation successfully. They recognize that a certain amount of post-editing (corrections made by a human translator) is necessary in order to produce production-quality translations. But in many cases, production quality is not required. The goal might be simply to give readers the gist of a document's content.

When implemented intelligently and used selectively, machine-translation software reduces translation costs substantially. Equally important, machine translation can make it possible to provide rough translations of information that, for economic reasons, otherwise could not be translated at all.

For an excellent overview of machine translation, see Dillinger and Lommel (2004).

Translation Memory

Virtually all technical translators use computer-assisted translation tools. One of the main components of these tools is translation memory (TM)—a database that stores the source-language version and the target-language version of every sentence that is translated. When a new or updated document is processed by the software, any translation segments that are identical or similar to previously translated segments are presented to the translator. The translator then decides whether to reuse, modify, or disregard the previous translations.

Unnecessary inconsistencies make the use of translation memory less efficient. For example, suppose that a French translator translates the following sentence in the first edition of a software manual:

- Use the Group column to see if your tables are joined in more than one group.

Later, a translator who is using the TM database from that first edition encounters the same sentence in a new version of the document or in a related document. Instead of retranslating the sentence, the translator can insert the previous translation with the click of a mouse or with a keyboard shortcut.

Now suppose that the writer or editor who worked on the second edition of the manual had decided to modify that sentence as follows:

- Use the Group column to **determine whether** your tables are joined in more than one group.

In this scenario, the translation-memory software finds the translation of the original sentence in its database and presents that sentence and its translation to the translator as a fuzzy match. However, now the translator has to decide whether the previous translation is suitable or whether it needs to be modified. Obviously, that task is more cognitively demanding and more time-consuming than inserting the translation of an exact match. When the English sentence is ambiguous or difficult to understand, the task is especially time-consuming, and the decision process is subject to error.

When you multiply the unnecessary variations in a document by the number of languages that the document will be translated into, the cost of those variations becomes very significant. Therefore, many of the Global English guidelines are aimed at eliminating sources of unnecessary variation.

Controlled-Authoring Software

Learning to follow all of the Global English guidelines could be a daunting task—although one could argue that it is no more daunting than learning the guidelines that are in any other style guide. However, there is one technology that greatly facilitates the task of following not only the Global English guidelines, but many other style guidelines as well. That technology is commonly referred to as controlled-authoring software.[3]

[3] The terms controlled-language checker and automated editing software are also used. However, the latter term sometimes refers to less sophisticated and non-customizable products that cannot provide adequate support for the Global English guidelines. See http://www.globalenglishstyle.com for a list of software vendors that license controlled-authoring software.

A controlled-authoring application parses texts and brings style errors, grammar errors, and terminology errors to the user's attention. One essential feature that distinguishes controlled-authoring software from other types of editing tools is that you can customize it. In collaboration with the software vendor, you can specify which grammar rules, style guidelines, and terminology restrictions you want the software to help authors follow.

You can also customize the rules to eliminate false alarms that are caused by idiosyncrasies in your documentation. Thus, the software is more accurate and reliable than off-the-shelf language checkers.

Many organizations use controlled-authoring software to ensure a high degree of language quality and consistency in their publications; to increase the productivity of content authors, editors, and translators; to help non-native authors produce better-quality English source texts; or for other business reasons.

Case Study

At SAS Institute, the implementation of controlled authoring was motivated partly by the need to standardize and control terminology. In recent years, SAS software products have become more integrated. SAS also began publishing documentation on the Web, with a consolidated index and full-text search. Terminology issues became more visible, both internally and to customers, than ever before.

The intensified pace of globalization also meant that SAS needed to find an efficient way of making its documentation more suitable for translation and easier for non-native speakers of English to understand. An earlier version of the Global English guidelines was developed for that reason and became an official part of the *SAS Style Guide for User Documentation*.

But even the best technical writers find it difficult to apply complex style guidelines or to consistently conform to lists of approved and deprecated terms. Deadlines and time pressures make it impractical for authors and editors to refer to style guides and glossaries frequently.

To emphasize the goal of helping authors communicate clearly and consistently, SAS used Assisted Writing and Editing (AWE) as the name of the project that encompassed the use of controlled-authoring software. After selecting a controlled-authoring product, SAS worked with the vendor to make the software as accurate as possible. For example, the software initially flagged the following sentence as an error because *the at* seemed to be an ungrammatical sequence of words:

- ▶ The remaining seven characters can include letters, digits, underscores, the dollar sign ($), or **the at** sign (@).

That false alarm was eliminated by modifying the rule so that it ignores any occurrence of *the at* that is immediately followed by *sign*.

In the following sentence, the controlled-authoring software initially flagged *a HMDA* as an error and suggested *an HMDA* instead:

- To view **a HMDA** Edit Analysis Report, complete these steps:

But *HMDA* is pronounced as an acronym (HUM-dah), not as an initialism (H-M-D-A). Therefore, *a HMDA* was added to an exclusion list so that it would no longer be flagged as an error.

Controlled-authoring software gives authors immediate feedback on their own writing, teaching them to follow guidelines that they might otherwise have difficulty understanding or remembering. SAS has found that after an initial productivity hit, this training effect leads to the opposite: a significant productivity *increase*. Because authors fix grammar, spelling, style, and terminology issues early in the writing process, there are fewer corrections to be made late in the documentation cycle, when the pressure to deliver is greatest.

The software's consistent, objective feedback reduces unnecessary variation. SAS anticipates that the increased consistency in its documentation will make the use of translation memory more effective, and that consistent terminology and phrasing will make its documentation more usable for all audiences.

SAS is working closely with the vendor to further customize the software so that it will detect violations of more of the Global English guidelines. The software already detects violations of most of the other style guidelines and terminology restrictions in the *SAS Style Guide*.

For more details about implementations of controlled-authoring software, see Akis and Simpson (2002) and Kohl (2007).

Practical Considerations for Implementing Global English

Prioritize the Guidelines

Whether you use controlled-authoring software or not, you will probably want to focus on a subset of the Global English guidelines first. To help you decide which guidelines are most important for your circumstances, the heading for each major style guideline is followed by a Priority line that looks like this:

Priority: HT1, NN2, MT3

The following tables explain the acronyms and priority levels:

Acronym	Meaning
HT	human translation
NN	non-native speakers
MT	machine translation

Priority Level	Meaning
1	high priority
2	medium priority
3	low priority

For example, HT1 indicates that the guideline has high priority for documents that will be translated by human translators. NN2 indicates that the guideline has medium priority for untranslated documents that will be read in English by non-native speakers. And MT3 indicates that the guideline has low priority for documents that will be translated using machine-translation software. These priority values are based on the author's subjective assessments and on feedback from translators and other localization professionals.

In Appendix B, "Prioritizing the Global English Guidelines," the style guidelines are presented in tables that are sorted by the HT, NN, and MT values.

Build a Relationship with Your Localization Staff

If you don't already know who manages the localization of your products and documentation, find out! Let them know that you are working toward making the localization process more efficient by improving the quality of your English documentation. Open a communication channel so that you will have someone to turn to when you have questions about whether a particular issue poses a problem for translation or localization. You, in turn, can point them to the right person if they have a question about the content of a particular document or product.

Always provide a glossary to your localization coordinator before the localization process begins. For more information about what to include in the glossary, see guideline 3.7.1, "Consider defining or explaining noun phrases."

Eliminate Non-essential Information

In addition to following the Global English guidelines, be sure to consider other ways of reducing translation costs. One of the best ways is to reduce the volume of information to be translated. Content reduction can be done at the topic level, at the sentence level, or both.

Topic-Level Content Reduction

Many technical documents contain topics that are of interest to only a small percentage of readers. For example, as part of a "Downsizing Our Documentation" campaign at SAS Institute, a team of technical writers, software developers, and technical support analysts was able to remove 30% of the content of a 500-page technical reference manual. Many topics in the document were of interest primarily to the technical support analysts and were therefore relocated to an internal Web site that is not translated.

According to one of the software developers, some of the information was there "to fill the term paper requirement." That is, the corporate culture seemed to require that if you developed new functionality, the functionality had to be documented in the user manual, even if it was of interest to only a few customers. Obviously, a new feature that "requires" twelve pages of documentation seems more impressive than a new feature that "requires" only six. No one on the downsizing team was previously aware that the document was being translated into six languages at an average cost of $.25 per word for each of those languages.

The technical writer was able to eliminate an additional 10% of the content by improving the organization of the document. He consolidated topics that were addressed in multiple places, and he was able to eliminate unnecessary introductions by using more-descriptive headings.

The difficulty with the topic-level approach is that it can require a considerable amount of time, effort, and coordination. Many organizations are not committed enough to the goal of reducing localization costs to assemble a team that has the right qualifications for deciding which content can be removed. Other priorities take precedence—especially since the division that pays for localization (and which would therefore reap the benefits of content reduction) is usually separate from the division that produces the documentation.

Sentence-Level Content Reduction

As the examples in Appendix A, "Examples of Content Reduction," illustrate, even essential topics can usually be shortened by removing unnecessary sentences and by making remaining sentences more concise.

Unlike topic-level content reduction, sentence-level content reduction can be done by an individual author or editor, or by a team of authors and editors. The participation of subject-matter experts from other divisions is not required. The advantage of a team is that team members develop shared strategies for reducing content, which can then be applied to many documents.

With practice, it is possible to focus on the Global English guidelines and on sentence-level content reduction at almost the same time. Throughout this book, you are frequently encouraged to find a more concise way of expressing an idea instead of merely applying a Global English guideline.

For more information about content reduction, see Rushanan (2007) and Fenstermacher (2006).

Don't Eliminate Syntactic Cues

Even though you should always look for opportunities to be more concise, don't remove syntactic cues from your documents. Syntactic cues are function words, punctuation marks, and other language features that are optional in some contexts. For example, in the sentence below, the word *that* is a syntactic cue. It can be omitted without making the sentence ungrammatical, but its presence makes the sentence structure more explicit.

▸ Ensure **that** the power switch is turned off.

Many of the Global English guidelines encourage you to use syntactic cues in order to eliminate ambiguities and to improve the readability of your sentences. Therefore, syntactic cues should not be removed in order to reduce word counts.

Syntactic cues are discussed in detail in Chapter 6, "Using Syntactic Cues," and in Appendix D, "Improving Translatability and Readability with Syntactic Cues."

Insert Explanations for Translators

As you will see when you read other parts of this book, sometimes a clause or sentence is ambiguous and there is no practical way to make it unambiguous. It is best to prepare for that situation by having a standard way of inserting explanations into your text for your translators' benefit.

Here is an example. In the following sentence, it is not entirely clear whether the relative clause, *that contains the data source*, modifies *Folders tree* or *location*:

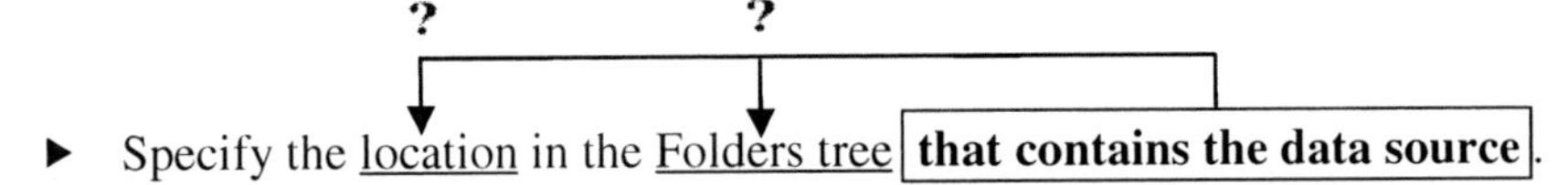

In other words, does the location contain the data source, or does the Folders tree contain the data source?

Suppose the author knows that the relative clause modifies *location*. Usually it is better to place a relative clause as close as possible to what it is modifying:

▸ Specify the location **that contains the data source** in the Folders tree.

But the above revision causes a different ambiguity. Now a translator might misinterpret the prepositional phrase, *in the Folders tree*, as modifying *data source*.

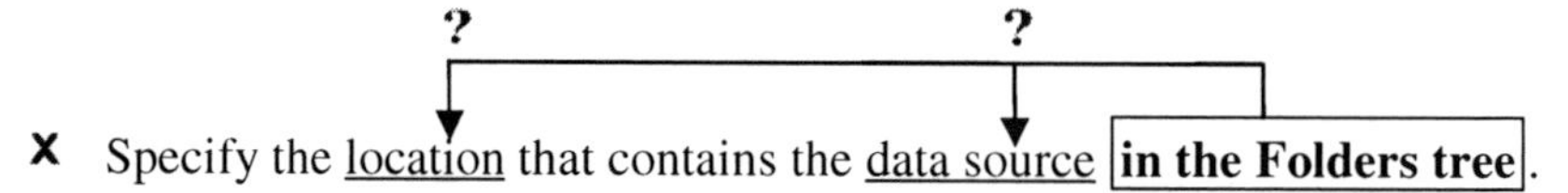

In fact, it, too, modifies *location*. That is, the location contains the data source, and the name of the location is displayed graphically in a tree-like, hierarchical list of folders.

For XML documents, the Internationalization Tag Set (ITS), Version 1.0, includes a Localization Note data category that is used for this purpose. (See http://www.w3.org/International/its.) You can use the <locNote> tag to include localization notes for your translators, as in the following example:

▶ Specify the location in the Folders tree that contains the data source.
```
<locNote>"that contains the data source" modifies
"location"</locNote>
```

Whatever publishing tool you use, it should provide some way of inserting comments that are visible only when another user of those tools specifies a particular setting or option. Thus, you can provide these explanations to translators without including the explanations in your deliverables.

For more examples of contexts in which you might need to provide explanations to translators, see guidelines 4.2.5, 4.6.7, 5.1.1, and 5.1.2.

Frequently Asked Questions about Global English

What is the relationship between Global English and controlled English?

First, controlled English is not a single entity. The term describes any of several attempts to define a subset of the English language that is simpler and clearer than unrestricted English. Most versions of controlled English specify which grammatical structures are allowed and which terms are allowed, as well as how those terms may be used. In early forms of controlled English, terminology was often restricted to a core vocabulary (in some cases, as few as 800–1000 terms), supplemented by technical terms that are necessary for a particular subject area or product.

Global English could be regarded as a loosely controlled language, yet it was developed using almost the opposite approach. In the development of Global English, the emphasis has been on identifying grammatical structures and terms that should be *avoided*, rather than on cataloging all of the grammatical structures and terms that are *allowed*. In other words, anything that is not specifically prohibited or cautioned against is allowed.

When texts conform to the guidelines and terminology restrictions of the more restrictive forms of controlled English, the style and rhythm of those texts differs noticeably from the style and rhythm of unrestricted technical English. By contrast, most readers don't notice anything different about the style and rhythm of texts that conform to the Global English guidelines.

Early versions of controlled English, such as the Kodak International Service Language (KISL), were developed as alternatives to translation. By severely limiting the range of grammatical structures and vocabulary that are allowed, KISL makes technical documents understandable even to readers who have very limited proficiency in English. Kodak found that it is much less expensive to teach service technicians all over the world a limited amount of English than to translate service manuals into 40 or more languages.

Global English is an alternative to translation only if the non-native speakers in your audience are reasonably proficient in English. Global English can make the difference between documents that those non-native speakers can read easily and documents that are too difficult for that audience to comprehend.

If you are writing for readers who have limited proficiency in English, then consider using a form of controlled English. However, keep in mind that the amount of effort and knowledge that is required for developing and implementing controlled English is considerable. Consult the Bibliography section of this book for sources of more information about controlled English.

Do the Global English guidelines make all sentences clear and easy to translate?

As Farrington (1996) said, "Simplified English will not compensate for a lack of writing skills." He was referring to AECMA Simplified English, the forerunner to ASD Simplified Technical English,[4] but the same could be said of Global English. Authors who follow the Global English guidelines still have to have good writing skills in order to produce high-quality prose.

Does following these guidelines lead to an increase in word counts?

According to Bernth (1998b), "In order to cut down on ambiguity, it is nearly always necessary to be somewhat verbose." However, the guidelines in this book include frequent reminders to search for opportunities to be more concise. Authors who pay attention to those reminders are able to eliminate unnecessary text and verbiage while they are applying the Global English guidelines.

Even if your word counts increase slightly, the benefits of increased clarity, readability, and consistency outweigh the additional per-word cost of translation.

---

[4] ASD Simplified Technical English is a form of controlled English that was initially developed for use in the aerospace industry. It is now used in other industries as well.

Typographical Conventions

This book contains hundreds of example sentences. The following symbols indicate what each example is intended to illustrate or represent:

Symbol	Meaning
X	An example that, from a Global English perspective, is incorrect or undesirable.
✓	A revised example that conforms to the Global English guideline that is being presented.
✓+	A revised example that conforms to the Global English guideline that is being presented and that is stylistically better than a previous revision.
?	A revised example that might be acceptable to some people but not to others, or that might be technically incorrect.
⊘	An example that conforms to Global English guidelines but that is either technically incorrect or stylistically unacceptable.
▶	An example that illustrates a point of grammar or style rather than illustrating compliance or non-compliance with a Global English guideline.
T	A translation of an example sentence.

Chapter 2

Conforming to Standard English

Introduction

As noted in the "Translation Memory" section of Chapter 1, unnecessary variations in phrasing and terminology make the translation process inefficient. This chapter focuses on unnecessary variations that result from non-standard uses of English. For example, in the following sentence, the term *modern-like* undoubtedly seems strange to most native speakers of English:

▸ Not all scientists agree that **modern-like** birds lived during the age of the dinosaurs.

In most contexts, the suffix *-like* is used only with nouns (*life-like*, *god-like*, *claw-like*, and so on), never with adjectives. Thus, except in a few subject areas such as paleontology and anthropology, *modern-like* is very "strange-like" indeed! The term is unlikely to be found in dictionaries, and the chances of its being in a machine-translation lexicon or in a translator's terminology database are very slim.

As the above example illustrates, usage that is considered non-standard in most subject areas is sometimes considered standard in other areas. But you probably have a sense of what is common and accepted in your particular subject area. When you encounter a linguistic phenomenon that sounds odd, consult your colleagues or refer to the resources that are recommended at the end of this chapter.

2.1 Be logical, literal, and precise in your use of language

Priority: HT1, NN1, MT1

Pay close attention to the literal meaning of each sentence that you write. If you express an idea that is illogical or imprecise, translators have to spend extra time correcting your errors.

Translators naturally pay very close attention to literal meanings and are very likely to notice the problem in the following sentence:

✗ This report compares **the salaries of different departments** for employees who have the same education level.

This sentence is illogical because departments don't earn salaries—employees do. What the author meant was this:

✓ This report compares **the salaries of employees** who have the same education level, grouped by department.

Aside from the extra burden that such errors place on human translators, poor word choice and faulty logic also have consequences for machine translation. Machine-translation software translates everything literally. Literal translations of sentences like the following often sound ridiculous even though many readers are not disturbed by the original English:

> ✗ When **you hover** over a menu item, a white underline appears.

Even a human translator might find it difficult and time-consuming to find an acceptable way of translating this sentence. Here is a logical, literal way of expressing the same idea:

> ✓ When **you position your mouse pointer** over a menu item, a white underline appears.

2.2 Use nouns as nouns, verbs as verbs, and so on

Priority: HT2, NN2, MT1

Use common words only as they are classified and defined in standard dictionaries.[1] For example, since *action* is listed in standard dictionaries only as a noun, you should not use it as a verb:

> ✗ The last recipient did not approve the request in the allotted time. In order to **action** this report, you must respond by selecting either `Resend to Approver` or `Resend to Approver's Manager`.

In most other languages, a word usually cannot be used as more than one part of speech: The noun equivalent for *action* cannot simply be converted to a verb. Therefore, translators must find some other way of translating this word. Such decisions slow down the translation process and can lead to errors in translation. (Imagine that you are a translator. Is the meaning of *action* in the above sentence clear to you?)

[1] As explained in "Researching Orthographic Variants" in Chapter 9, standard dictionaries are not the best or only sources to consult for guidance on hyphenation or orthography. However, it is certainly appropriate to consult standard dictionaries for information about word meanings, parts of speech, and transitivity.

Another problem is that *action* won't be recognized as a verb by machine-translation software unless it has been specifically identified as a verb in the software's lexicon. This non-standard usage will confound the software's analysis and will likely cause other parts of the introductory phrase to be mistranslated as well. Here is an appropriate revision:

✓ The last recipient did not approve the request in the allotted time. To **resubmit** the report **for approval**, select either `Resend to Approver` or `Resend to Approver's Manager`.

Some authors object to the use of *partner* as a verb, as in this example:

✓ We **partner** with our customers to turn data into usable knowledge.

However, according to the *Merriam-Webster Online Dictionary*, *partner* has been used as a verb since 1611, and translators can easily understand what *partner* means in this context.

By contrast, most dictionaries define *mouse* as a verb only in the sense of *to hunt for mice*. Therefore, the following usage is unacceptable:

X **Mouse** back up to the `CS Variable` field.

The sentence can easily be revised as follows:

✓ **Move the mouse pointer** back up to the `CS Variable` field.

Here are some additional examples of words that have been forced into non-standard roles:

A verb used as a noun

Import can be used as a noun in some contexts. However, instead of using a noun to express the action in a clause or sentence, turn the noun into a verb whenever possible:

X The **import** of the data into MySQL is also very simple.

✓ It is also very easy to **import** the data into MySQL.

See also guideline 3.3, "Use a verb-centered writing style."

An adjective used as a noun

By omitting the noun in a noun phrase, authors frequently force adjectives into the role of nouns. This practice poses a serious problem for human translators and for machine-translation systems.

Consider the following sentence, in which the author used the adjective *pop-up* as a noun:

X This section explains how to change the text of items in the **pop-up**.

In many languages, nouns have gender (typically masculine, feminine, and sometimes neuter). An adjective such as *pop-up* has different forms, depending on the gender of the noun that it modifies. Therefore, translators need to know what noun *pop-up* modifies in order to translate it. The author should use the complete noun phrase, *pop-up menu*, instead of using *pop-up* as an adjective:

✓ This section explains how to change the text of items in the **pop-up menu**.

An adjective used as a verb

X EN-15038 will **obsolete** all other standards for assessing translation quality.

✓ EN-15038 will **make** all other standards for assessing translation quality **obsolete**.

X Whenever you **later** a defect, be sure to **later** it to the next production cycle.

✓ Whenever you **select LATER** (to defer the defect), be sure to assign the defect to the next production cycle.

2.3 Don't add verb suffixes or prefixes to nouns, acronyms, initialisms, or conjunctions

Priority: HT1, NN3, MT1

In most languages, you cannot convert a noun, acronym, initialism, or conjunction to a verb by adding a verb suffix or prefix to it. Therefore, this type of word typically must be translated as a phrase, not as a single word. It is difficult and time-consuming for translators to determine what such words mean and to find other ways of expressing them. It is also difficult to customize machine-translation systems to handle such words appropriately.

Nouns

In the following sentence, the author has used the noun VDEFINE (the name of a program statement) as a verb by adding the past-tense suffix *-d*:

X If a variable is **VDEFINEd** more than once in any step, then the next reference to that variable will cause a storage overlay.

The sentence could be rephrased as follows:

✓ If a variable is **defined by more than one VDEFINE statement** in any step, then the next reference to that variable will cause a storage overlay.

Acronyms

In the following example, the acronym RIF (reduction in force) has been converted to a verb:

✗ Norlina Industries reported yesterday that it has **riffed** 45 of its 900 employees.

✓ Norlina Industries reported yesterday that it has **laid off** 45 of its 900 employees.

Conjunctions

✗ Any value that follows a plus sign is **OR'ed** with the current option value.

✓ Any value that follows a plus sign is **combined** with the current option value **in a Boolean OR expression**.

2.4 Use standard verb complements

Priority: HT2, NN2, MT1

With one exception (explained in guideline 2.5), standard dictionaries don't tell you which grammatical constructions are typically used as complements for common verbs. However, if you are not sure whether a particular verb construction is considered standard or acceptable by most native speakers of English, you can consult the references listed in "Useful Resources" on page 30.

Here is an example of a non-standard verb complement:

✗ We **recommend** to use the Web version of the application.

The use of an infinitive complement (*to use . . .*) with the verb *recommend* is unnatural to most native speakers of English. This non-standard usage is likely to cause problems for machine-translation systems, and it is an unnecessary variation of the following, more acceptable revision:

✓ We **recommend** that you use the Web version of the application.

Instead of changing a verb's complement, sometimes it is better to choose a different verb:

✗ By default, a command bar is displayed at the top of the window. Alternatively, you can **select** to display a floating command dialog box instead.

✓ By default, a command bar is displayed at the top of the window. Alternatively, you can **choose** to display a floating command dialog box instead.

Also consider more-drastic revisions such as the following:

- ✓ Alternatively, you can **display** a floating command dialog box by selecting `Command box` from the Preferences dialog box.

More about Verb Complements

A *complement* is a word or phrase that completes a grammatical construction. A *verb complement* completes the predicate of a sentence. Most of the time, native speakers of English use standard verb complements without even knowing what verb complements are.

Here are some examples of different types of complements that the verb *start* can take:

- ✓ Suddenly, the aliens **started** to speak. (infinitive)
- ✓ Suddenly, the aliens **started** speaking. (gerund)

Here are some complements for the verb *become*:

- ✓ Lola **became** smug. (predicate adjective)
- ✓ Lola **became** an engineer. (predicate noun)

Many verbs can take more than one type of complement, but in some cases the meaning of the verb differs when different complements are used:

- ✓ Ali **ran** to the store. (prepositional phrase)
- ✓ James **ran** a high-speed printing press. (direct object)
- ✓ The mad scientist **initiated** the destruct sequence. (direct object)
- ✓ The professor **initiated** Jason into the mysteries of psycholinguistics. (direct object + prepositional phrases)

However, the following combinations of verbs and complements are clearly ungrammatical:

- X Suddenly, the aliens **started** that they spoke. (noun clause)
- X Lola **became** screaming. (gerund)

2.5 Don't use transitive verbs intransitively, or vice versa

Priority: HT3, NN3, MT1

Transitivity is a type of verb complementation that is familiar to many authors because it is included in standard dictionary entries. A *transitive verb* has a direct object as a complement; an *intransitive verb* has no direct object. You can consult standard dictionaries to determine whether a verb is typically used transitively, intransitively, or both.

As you saw in guideline 2.4, a verb often has different meanings depending on whether it is used transitively or intransitively. The meaning of the verb *ran* is different in *Ali ran to the store* than it is in *James ran a high-speed printing press*. In the first sentence, *ran* is intransitive because it has a prepositional phrase (*to the store*) as a complement, but no direct object. In the second sentence, *ran* is transitive because *a high-speed printing press* is a direct object.

According to the *Merriam-Webster Online Dictionary*, when the verb *display* is used intransitively, its only conventional meanings are *to show off* or *to make a breeding display*. An example would be *The penguins displayed in an elaborate mating ritual.* Therefore, the following sentence is non-standard:

✘ If the wrong video driver has been installed, your program might **display** strangely.

You can resolve the problem by using passive voice:[2]

✔ If the wrong video driver has been installed, your program might **be displayed** strangely.

Using the intransitive verb *rise* with a direct object (*the management ladder*) is also non-standard and is quite disturbing to most native speakers:

✘ Davis tells students not to shun jobs in restaurants, because restaurant employees can often **rise** the management ladder quickly.

The author should have used the verb *climb* instead:

✔ Davis tells students not to shun jobs in restaurants, because restaurant employees can often **climb** the management ladder quickly.

In order to further reduce variation in your documentation, you might consider mandating that a verb be used only transitively or only intransitively even if

---

[2] Because the performer of the action is unimportant, the use of passive voice is acceptable. See guideline 3.6, "Limit your use of passive voice," for further discussion of contexts in which passive voice is appropriate.

dictionaries list both usages. For example, it's very unusual to use the verb *pause* with a direct object, as was done in this sentence:

X If you are not sure what an icon represents, **pause** your cursor on the icon.

In this case, instead of using passive voice, you can use a different verb and preposition:

✓ If you are not sure what an icon represents, **position** your cursor **over** the icon.

⚠ An Exception to This Guideline

Sometimes a verb that was formerly used only transitively comes into common usage as an intransitive verb, or vice versa. In such cases, consider accepting the change in usage.

For example, the verb *shut down* is commonly used with a direct object, as in *Shut down your computer before you leave for the weekend*, where *your computer* is the direct object. At one time, intransitive usages of this verb were unusual. However, the intransitive usage has become so common that the transitive versions of the following sentence sound odd:

✓ Please wait while your computer **shuts down**. (intransitive)

? Please wait while the operating system **shuts down** your computer. (transitive, active)

? Please wait while your computer **is shut down**. (transitive, passive)

Here is another example of intransitive usage that has become common in software documentation:

✓ The statements are executed immediately after the software **has** fully **initialized**. (intransitive)

? The statements are executed immediately after the software **has been** fully **initialized**. (transitive, passive)

2.6 Use conventional word combinations and phrases

Priority: HT3, NN3, MT1

Native speakers instinctively associate certain nouns and prepositions with certain verbs, certain adjectives with certain nouns, and so on. For example, the verb *correspond* is typically used with the prepositions *to* and *with*, the adverb *guardedly* is almost always followed by the adjective *optimistic,* and the noun *felony* is most frequently used with the verb *commit*.

Unusual word combinations, such as the use of the preposition *to* with the verb *associate*, disturb native speakers:

> **X** The user can **associate** metadata **to** any metadata resource in a repository.

The author should have used *with* instead:

> ✓ The user can **associate** metadata **with** any metadata resource in a repository.

Unusual word combinations and phrases also increase translation costs in the following ways:

- Experienced human translators are accustomed to using standard translations for standard word combinations and phrases. Unusual word combinations and phrases require translators to give more thought to the intended meaning.
- Machine-translation software typically looks at the context to determine how to translate a particular word. When you use unusual word combinations and phrases, you decrease the likelihood that the software will produce a correct translation.
- Non-standard word combinations and phrases reduce consistency in documentation, which makes the use of translation memory less efficient.

If you are not sure whether a particular combination of words is conventional or not, consult your co-workers or consult the references listed in "Useful Resources" on page 30.

2.7 Don't use non-standard comparative and superlative adjectives

Priority level: HT3, NN3, MT3

Don't use *more* and *most* to form the comparative and superlative of adjectives for which *-er* and *-est* endings are much more common. You can use a Web search to test your assumptions about which comparatives and superlatives are more common. Include some of the context to ensure that your results are not skewed by contextual differences. For example, a Google search of *times narrower* returned 35 times as many hits as *times more narrow*:

- ✗ Can particles that are tens of thousands of times **more narrow** than a human hair penetrate human skin?
- ✓ Can particles that are tens of thousands of times **narrower** than a human hair penetrate human skin?

Similarly, *one of the most common* greatly outnumbers *one of the commonest* (in U.S. English, at least), and *are more likely to* greatly outnumbers *are likelier to*:

- ✗ Gene-chip technology can detect one of the **commonest** genetic mutations with 100% accuracy.
- ✓ Gene-chip technology can detect one of the **most common** genetic mutations with 100% accuracy.
- ✗ Ants are **likelier** to take bait when the temperature is 65 to 85 degrees Fahrenheit and when the ground is dry.
- ✓ Ants are **more likely** to take bait when the temperature is 65 to 85 degrees Fahrenheit and when the ground is dry.

2.8 Use *the* only with definite nouns

Priority: HT3, NN3, MT3

The article *the* occurs more frequently than any other word in the English language, yet even professional writers and editors often use it incorrectly. The incorrect usage confuses readers who are trying to understand unfamiliar material. Because the incorrect use of *the* is often translated into incorrect use of the corresponding article (if there is one[3]) in the target languages, this issue is significant for translated documents as well as for untranslated documents.

When you use the definite article *the* plus a noun, you imply that your reader knows specifically which instance of the noun you are referring to. If you have not mentioned the noun before, and if there is no other way for readers to understand which particular instance of the noun you are referring to, then your readers will be confused.

For example, suppose someone says to you *I saw* **the** *dog on my way to work*. That statement implies that you and the speaker have talked about this dog before, and that you therefore know specifically which dog the speaker is referring to. If you and the speaker have not discussed the dog before, then the speaker's use of *the* seems puzzling.

In the following example, the use of *the* with *process* in the second sentence is confusing because there has been no previous mention of a process:

> ✗ The DBE option specifies how many disk blocks RMS adds when it automatically increases the size of a table during a Write operation. If you specify DBE=0, then RMS uses **the process'** default value.

In order to make it clear which process is being referred to, the author must be more explicit, as in the following revision of the sentence:

> ✓ If you specify DBE=0, then RMS uses the default value for **the process that is performing the Write operation**.

In the following definition, the use of *the hierarchy* is confusing. Nowhere on the page that this definition appeared on was a hierarchy mentioned, so it is not clear which hierarchy is being referred to:

> ✗ HIERLIST is a list that shows **the hierarchy**.

The author should use *a hierarchy* instead or else should add some modification (for example, *the hierarchy that was specified for the organizational chart*).

For a better understanding of the complexities of article usage in English, see Kohl (1990).

---

[3] Many languages (Japanese, Korean, and Chinese, for example) don't have equivalents for *a*, *an*, or *the*. For information about article usage in other languages, see Swan, Smith, and Ur (2001).

2.9 Use singular and plural nouns correctly

Priority: HT3, NN3, MT3

This guideline is related to guideline 2.1, "Be logical, literal, and precise in your use of language." When you are discussing two nouns that exist in some kind of a relationship, be sure to convey the relationship accurately. When you use a singular noun where you should have used a plural noun (or vice versa), you distort your meaning.

For example, in the following sentence, it is illogical and incorrect to say that all data items (plural) have a (single) numeric value, unless they all have the same numeric value:

X All the data items shown in Figure 7 have a numeric value.

If the data items all have the same value, then state that idea explicitly:

✓ All **the data items** shown in Figure 7 **have the same numeric value**.

If the data items all have multiple numeric values, then this version of the sentence is correct:

✓ All **the data items** shown in Figure 7 **have numeric values**.

If each data item has a single numeric value, then this version of the sentence is correct:

✓ **Each data item** shown in Figure 7 **has a numeric value**.

If each data item has a unique numeric value, then this version of the sentence is correct:

✓ **Each data item** shown in Figure 7 **has a unique numeric value**.

Translators and machine-translation systems can easily translate any of the above sentences. However, one of the most important underlying principles of Global English is to use the English language precisely. If a sentence is incorrect in English, then of course the translations will also be incorrect.

The next sentence incorrectly states that column names (plural) can be qualified with the name of only one particular table:

X **Column names** can be qualified with **a table name**.

Either of the following revisions would be acceptable:

✓ **A column name** can be qualified with **a table name**.

✓ **Column names** can be qualified with **table names**.

This next sentence incorrectly states that all new files are created by only one person:

✗ Enter `x umask 022` to ensure that **new files** can be written to only by **their owner**.

Yet if you say *owners*, the sentence is ambiguous, because it could mean that a single file could have more than one owner. In this case, that is not true. Here is a revision that solves the problem:

✓ Enter `x umask 022` to ensure that **each new file** can be written to only by **its owner**.

Other Guidelines That Pertain to Standard English

The following guidelines also pertain to the use of standard English but are discussed in other chapters:

- 6.2, "In a series of noun phrases, consider including an article in each noun phrase"
- 7.3, "Revise dangling -ING phrases"
- 7.4, "Punctuate -ING phrases correctly"
- 8.13.3, "Don't capitalize common nouns"

Useful Resources

As you might expect, native speakers of English don't always agree on which complements particular verbs can take, on whether particular verbs are transitive or intransitive, or on which word combinations are standard or non-standard. To help resolve any disagreements, consult the following sources:

- Benson, Morton, Evelyn Benson, and Robert Ilson. 1997. *The BBI Dictionary of English Word Combinations.* Revised edition. Amsterdam and Philadelphia: John Benjamins.
- Hornby, A.S., and Sally Wehmeier. 2005. *The Oxford Advanced Learner's Dictionary of Current English.* 7th edition. London: Oxford UP.

Chapter 3

Simplifying Your Writing Style

Introduction

If you have studied a foreign language, then you know that your ability to read and comprehend a foreign-language text depends not only on the vocabulary, but also on the complexity of the author's writing style. Even when you read a text that is written in your native language, style matters at least as much as vocabulary.

If you are a native speaker of English and have *not* studied a foreign language, you probably don't fully appreciate the challenge that unnecessary complexities pose for non-native speakers and translators. Languages differ from each other more than you might realize, making the reading and translation processes much more than simple word-substitution exercises.

The following examples of English sentences that have been translated into other languages illustrate how different some languages are from English.[1] In addition to the differences in word order, notice that a single English word is often translated as multiple target-language words, and vice versa.

[1] The Japanese and Chinese examples are from Swan, Smith, and Ur (2001).

Example 1: Japanese

Japanese word order is strikingly different from word order in English. For example, prepositions (such as *in*, *on*, and *about*) follow their objects, subordinating conjunctions (such as *if*, *although*, and *when*) follow their clauses, and modal verbs (such as *can* and *might*) follow main verbs.

English:

I'm upset because people think I said something strange, when I said nothing at all.

Japanese, followed by a literal, word-for-word English translation:

```
Watakushi wa nani mo iwanakatta no ni,     hen na
I-as-for  what-also  said-not   although, strange-being

koto o yutta yoo ni omowarete     komarimashita.
thing  said  way-in thought-being troubled.
```

Example 2: Chinese

Not surprisingly, Chinese is also radically different from English.

English:

Thank you for your letter, and thanks for the regards.

Chinese, followed by a literal, word-for-word English translation:

```
lai  xin    shou dao le, xiexie ni de wen hou.
come letter receive,      thank  your  regard.
```

Example 3: German

Even though German and English are closely related, the differences in word order can still be significant.

English:

I didn't expect to be able to attend the conference.

German, followed by a literal, word-for-word English translation:

```
Ich erwartete nicht, dass ich die Tagung     hätte
I   expected   not,    that I  the conference  would have

beiwohnen können.
attend     be able to.
```

Obviously, you cannot change the sequence of words or ideas in English in order to accommodate readers from other language backgrounds. But if you express yourself simply and eliminate unnecessary obstacles, you can greatly facilitate the reading and translation processes.

3.1 Limit the length of sentences

Priority: HT1, NN1, MT1

Short sentences are less likely than long sentences to contain ambiguities and complexities that impede translation and reduce readability. For procedural (task-oriented) information, limit your sentences to 20 words. For conceptual information, strive for a 25-word limit.

Here are two examples of long sentences that can easily be divided into shorter sentences:

✗ If Chocolate Bits is set to `No`, indicating that there are no chocolate bits in the sample batch of ice cream, then the selections for Enough Bits and Size of Bits are grayed to prevent users from entering irrelevant data. (40 words)

✓ If Chocolate Bits is set to `No`, then there are no chocolate bits in the sample batch of ice cream. **Therefore**, the selections for Enough Bits and Size of Bits are grayed to prevent users from entering irrelevant data. (20 words + 19 words = 39 words)

✗ With design-time controls, you control the look and feel of your Web pages in a WYSIWYG editor environment, and at the same time use all the functionality of SAS/IntrNet software in your Web pages. (35 words)

✓ With design-time controls, you control the look and feel of your Web pages in a WYSIWYG editor environment. **In addition, you can** use all the functionality of SAS/IntrNet software in your Web pages. (19 +15 = 34 words)

Some long sentences are very difficult to divide or shorten. In the following example, the technical writer who received the sentence from a subject-matter expert had to study the context for half an hour. Only then did she understand the sentence well enough to divide and simplify it.

> ✗ The log shows the `Uninitialized variable` warning for any variable whose value comes from ISPF services when the variable has no initial value and does not appear on the left side of the equal sign in an assignment statement. (39 words)

Sentences like this one pose the biggest problem for translators. If the document in which the above sentence appears must be translated into ten languages, then ten different translators must struggle to understand the sentence. What are the chances of them all producing clear, accurate translations when they are given such impenetrable source material? If the subject matter were nuclear reactors instead of software, unclear source texts and mistranslations could be catastrophic.

Fortunately, the writer used her understanding of the subject matter to completely reorganize and clarify the information:

> ✓ If a variable has no initial value and does not appear on the left side of the equal sign in an assignment statement, then the ISPF service assigns a value to the variable. However, because the value was not assigned in a statement, the log shows the `Uninitialized variable` warning. (33 + 17 = 50 words)

In the revision, the first sentence still exceeds the 25-word limit for conceptual information, but it is divided into two clauses, each of which is unambiguous. The second sentence adds two clauses (and 17 words), but it includes a useful added explanation, and the word *However* makes the logical relationship between the sentences clear. The total word count increased from 39 words to 50 words, but the revision was necessary in order to make the information comprehensible (even to native speakers) and translatable.

The revision illustrates the importance of being flexible with regard to sentence lengths. A long sentence might be clear and reasonably translatable if the following conditions are true:

- The sentence consists of more than one clause. (However, avoid including more than two clauses in a sentence.)
- Each clause conforms to all of the other Global English guidelines.
- The logical relationship between the clauses is clear.

3.2 Consider dividing shorter sentences

Priority: HT3, NN3, MT3

Even sentences that are shorter than the recommended limits might benefit from being divided. In the following glossary definition, the relative clause can easily be made into a separate sentence:

✗ CGI-based technology: a technology that is based on the Common Gateway Interface standard**, which enables applications to communicate with Web servers**. (19 words in the definition)

✓ CGI-based technology: a technology that is based on the Common Gateway Interface standard**. CGI enables applications to communicate with Web servers.** (11 + 8 = 19 words in the definition)

3.3 Use a verb-centered writing style

Priority: HT2, NN2, MT2

Use verbs to convey the most significant actions in your sentences. This guideline is not specific to Global English, but it is so important for clear, readable, translatable communication that it deserves special emphasis.

In the following example, when the noun *encryption* is changed to a verb, the verb (*enables*) can easily be eliminated:

✗ A spawner program **enables the encryption** of user IDs and passwords when they are passed through a network. (18 words)

✓ A spawner program **encrypts** user IDs and passwords when they are passed through a network. (15 words)

The revised sentence makes it clear that the spawner program does more than *enable* the encryption of user IDs and passwords: It is the true agent of the verb *encrypts*. In addition, the revised sentence contains fewer words and therefore costs less to translate.

If this type of revision is not feasible, then consider converting the noun part of a verb + noun construction to an infinitive. In the next example, the verb *enable* was retained, but the noun *creation* was changed to the infinitive *to create*:

- X The templates **enable the rapid creation of** a decision-support framework that addresses your organization's needs.
- ✓ The templates **enable you to rapidly create** a decision-support framework that addresses your organization's needs.

Here are some other examples:

- X VMDOFF specifies whether **metadata verification checking** is to be performed on the data model.
- ✓ VMDOFF specifies whether **to verify** the data model's metadata.
- X The first note in the log indicates the **creation** of the Passengers table.
- ✓ The first note in the log indicates **that** the Passengers table **was created**.
- X The **addition of non-linear load** at the point of measurement limits the voltage scale of the instrument without affecting the transient severity assessment.
- ✓ **By adding non-linear load** at the point of measurement, **you** limit the voltage scale of the instrument without affecting the transient severity assessment.

Sometimes a complete rewrite is necessary in order to eliminate a weak, noun-based construction:

- X SSPI **enables transparent authentication** for connections between computers.
- ✓ SSPI **enables users to connect** to other computers **without supplying their user IDs and passwords**.

The following table lists several common verb + noun combinations along with their more-concise, single-verb alternatives:

Verb + Noun Combination	Verb-centered Alternative
reach an agreement	agree
come to a conclusion	conclude
make a decision	decide
provide an explanation	explain
conduct an investigation	investigate

3.4 Keep phrasal verbs together

Priority level: HT3, NN3, MT1

Whenever possible, keep the parts of a phrasal verb together:

✗ **Turn** the zoom tool **off** by clicking the circle tool.

✓ **Turn off** the zoom tool by clicking the circle tool.

There are several reasons for following this guideline:

- Separated phrasal verbs confuse those non-native speakers who are not accustomed to them.
- Following this guideline reduces unnecessary inconsistency. (As explained in Chapters 1 and 2, unnecessary variations in phrasing and terminology make translation less efficient and more expensive.)
- Most native speakers agree that following this guideline improves the style and readability of a sentence.
- This practice is better for machine translation.

Here are some other examples in which separated phrasal verbs can easily be revised to eliminate the separation:

✗ You can enable or disable a cube, fine-tune the performance of the server, or gracefully **shut** the server **down**.

✓ You can enable or disable a cube, fine-tune the performance of the server, or gracefully **shut down** the server.

✗ When in doubt, **spell** the name of the unit **out.**

✓ When in doubt, **spell out** the name of the unit**.**

✗ When the user **brings** the page **up** for viewing, the browser loads the image.

✓ When the user **displays** the page, the browser loads the image.

Note that it is not always possible to follow this guideline. Here is an example:

▸ When you **move** a column **up** in the list box, it will appear farther to the left in the table.

⊘ When you **move up** a column in the list box, it will appear farther to the left in the table.

The revision is unacceptable because using *move up* to refer to the action of physically moving an object other than oneself is not standard English. Thus, most readers interpret the first clause in the revision as *When you shift your gaze upward by one column in the list box*. Only when they encounter *it* in the main clause do they realize that that interpretation is incorrect.

3.5 Use short, simple verb phrases

Priority: HT3, NN3, MT2

In other languages, verb tenses[2] are not always linked to time, and different languages use different tenses to express the same point or range on the time axis. For example, in English we use present perfect progressive (*have been living*) to express what a German conveys using simple present (*wohne*):

- English: I **have been living** in Berlin for 12 years.
- German: Ich **wohne** seit 12 Jahren in Berlin.

To improve readability and to avoid unnecessary complications in both human translation and machine translation, use the simplest tense that is appropriate for each context.

3.5.1 Avoid unnecessary future tenses

In many contexts, a future tense (including future passive, future perfect, and so on) is necessary and appropriate, as in this example:

✓ You cannot predict which record **will be deleted**, because the internal sort might place either record first.

But in many other contexts, present tense works just as well:

X When you develop your application, test different values to determine which values **will result** in the best performance.

✓ When you develop your application, test different values to determine which values **result** in the best performance.

[2] For the sake of simplicity, the term *tense* is used here to refer to what technically should be referred to as tense and *aspect*. For a detailed explanation of the tenses of English, see Celce-Murcia and Larsen-Freeman (1998). For explanations of how tense in other languages differs from tense in English, see Swan, Smith, and Ur (2001).

Future tenses usually include the auxiliary verb *will*. Therefore, you don't need to learn the names of all the future tenses in order to recognize them. Just look for *will*, and ask yourself whether it is necessary. The following examples show different ways of eliminating unnecessary future tenses:

Future → present active

- ✗ A SUMSIZE value that is greater than MEMSIZE **will have** no effect.
- ✓ A SUMSIZE value that is greater than MEMSIZE **has** no effect.

Future passive → present active

- ✗ If the STYLE= option is used in multiple SUM statements that affect the same location, then the value of the last SUM statement's STYLE= option **will be used**.
- ✓ If the STYLE= option is used in multiple SUM statements that affect the same location, then SAS **uses** the value of the last SUM statement's STYLE= option.

Future passive → present passive

- ✗ DHTML that is specific to Internet Explorer **will** not **be processed** correctly by Netscape.
- ✓ DHTML that is specific to Internet Explorer **is** not **processed** correctly by Netscape.

Future verb + noun construction → present, verb-centered construction

- ✗ Higher selective pressure **will cause** faster **convergence** of the genetic algorithm.
- ✓ Higher selective pressure **causes** the genetic algorithm **to converge** more rapidly.

3.5.2 Simplify other unnecessarily complex tenses

Extreme violations of this guideline like the following are easy to recognize, but they occur infrequently:

- ✗ Scrolling to the right **should not be being performed** by any REXX application.
- ✓ REXX applications **should not scroll** to the right.
- ✗ The panel members all agreed that there **had been and was going to continue to be** considerable change in the industry.
- ✓ The panel members all agreed that the industry **would continue to change**.

More often, you simply have to pay attention to the verb constructions that you use in your writing. Ask yourself whether you can simplify and shorten them. Here are some examples:

Modal verb + present perfect → modal verb + present

✗ Before taking this course, you **should have completed** the following courses:

✓ Before taking this course, **complete** the following courses:

Past perfect passive → past passive

✗ The DUP function accesses records that have the same key as the current record, provided that the current record **had been accessed** using the CALC function.

✓ The DUP function accesses records that have the same key as the current record, provided that the current record **was accessed** using the CALC function.

Past perfect (subjunctive mood), modal + present perfect → present, present

✗ If the LABEL option **had been specified** in the input statement, then it **would** not **have been** necessary to use the DSN option.

✓ If you **specify** the LABEL option in the input statement, then you **do** not **need** to use the DSN option.

3.6 Limit your use of passive voice

Priority: HT2, NN3, MT1

Many style guides and technical writing textbooks advise against using passive voice except when there is a good reason for using it (as discussed below). From a Global English perspective, there are three main reasons for avoiding it:

- Passive voice is used more in English than in some other languages. In order to avoid producing texts that sound unnatural, translators often have to convert passive constructions to active voice.
- Active-voice constructions are usually more concise than passive-voice constructions. Reduced word counts lower the cost of translation.
- Some languages have more than one way of translating passive verbs. It is not always easy for translators to decide how to translate passive voice. Different translators might choose different ways of translating the same passive-voice construction. As a result, in a project that involves more than one translator for the same target language, inconsistencies arise within the translated document.

Here are some examples in which you can easily change passive voice to active voice:

- ✗ Accessibility standards for electronic information technology **were adopted** by the U.S. government in 1973. (14 words)
- ✓ The U.S. government **adopted** accessibility standards for electronic information technology in 1973. (12 words)
- ✗ The Percent column displays the percentage of the pie chart that **is represented** by each slice. (16 words)
- ✓ The Percent column displays the percentage of the pie chart that each slice **represents**. (14 words)
- ✗ To access the results that **are returned** by the query, use standard JDBC syntax. (14 words)
- ✓ To access the results that the query **returns**, use standard JDBC syntax. (12 words)
- ✗ The following steps **should be performed** to modify the Initial Contact Date: (12 words)
- ✓ To modify the Initial Contact Date, **perform** the following steps: (10 words)
- ✓+ To modify the Initial Contact Date, **follow these** steps: (9 words)

When Is Passive Voice Appropriate?

Passive voice is appropriate when the agent of the action is unknown or unimportant. For example, consider the following passive-voice sentence:

- ▸ This output **was created** by appending HTML output to an existing HTML file.

Most likely, the author of the software manual in which the sentence appeared created the output that the sentence is referring to. However, that information is irrelevant, and using the pronoun *I* in the following active-voice revision is inappropriate in a technical document:

- ▸ **I created** this output by appending HTML output to an existing HTML file.

Similarly, in the following sentence, the software (or some component of the software) that is being documented presumably completed the operation and generated the warning or note:

- ✗ A negative value indicates that the operation **was completed**, but a warning or a note **was generated**.

In a conceptual (as opposed to task-based) software manual, the software is often the agent of almost every passive verb. Changing each passive verb to active voice and repeating the name of the software over and over is stylistically unacceptable:

⊘ A negative value indicates that **Oracle completed** the operation, but that **Oracle generated** a warning.

Passive voice is also useful for emphasizing the recipient of an action rather than the agent that is performing the action. For example, in the following sentence, the phrase *The shared images* is the focus of the sentence:

► The shared images **are provided** by your system administrator.

If you change *are provided* to *provides*, *Your system administrator* receives the focus:

? Your system administrator **provides** the shared images.

Such changes in emphasis are often unacceptable.

In short, if you eliminate all passive verb constructions from conceptual information, you will probably violate the cardinal rule of Global English:

> **The Cardinal Rule of Global English**
> Don't make any change that will sound unnatural to native speakers of English.

On the other hand, remember the corollary to the cardinal rule:

> **Corollary**
> There is almost always a natural-sounding alternative if you are creative enough (and if you have enough time) to find it!

Don't be too quick to leave passive verbs unchanged or to use the first revision that comes to mind. In the following example, the revision is confusing because *explicitly* could be modifying either *specify* or *define*:

► In a spreadsheet, all functional relationships between numbers **are defined** by the formulas that you specify explicitly.

? In a spreadsheet, the formulas that you specify explicitly **define** all functional relationships between numbers.

You can eliminate the ambiguity by omitting *explicitly*:

✓ In a spreadsheet, the formulas that you specify **define** all functional relationships between numbers.

If shifting the focus from *formulas* to the user (*you*) is acceptable, then you could use the following revision:

✓ In a spreadsheet, you specify formulas that **define** the functional relationships between numbers.

3.7 Consider defining, explaining, or revising noun phrases

Priority: HT1, NN2, MT1

Noun phrases pose more problems for translators and for non-expert readers (regardless of whether they are native or non-native speakers) than any other syntactic structure in English. The following overview will help you understand the source and the magnitude of the problem before you delve into guidelines 3.7.1–3.7.3.

An Overview of Noun Phrases

A noun phrase can be a single noun, or it can consist of a noun plus one or more preceding determiners[3], adjectives, and other nouns. For example, the following sentence includes four noun phrases:

- Fire destroyed the one-story building in a matter of minutes.

Most forms of controlled English suggest revising noun phrases that are more than three words long.[4] However, even a two- or three-word noun phrase can be unclear or ambiguous. For example, in the following sentences, someone who is not familiar with the subject matter cannot fully understand the two-word noun phrases, because each individual word has multiple possible meanings:

- If you haven't imported a filter, the default is a **unity gain**.
- The **tracking loop** mitigates the effects of multi-path interference on code-phase errors.

---

[3] *determiner*: a word that precedes a noun and that either quantifies or helps to identify the noun. In English, determiners include articles (*a*, *an*, and *the*), numbers, quantifiers (*many*, *much*, *some*, *several*, *a few*, and so on), demonstrative pronouns (*this*, *that*, *these*, and *those*), and possessive pronouns (*my*, *your*, *their*, and so on).

[4] In this book, word counts that are given for noun phrases do not include articles or other determiners. However, word counts for sentences include all words.

On the other hand, some longer noun phrases are easy to comprehend—especially if part of the noun phrase is a proper noun. As long as the reader understands the two-word noun phrase *dialog box*, the four-word noun phrase in the following sentence is comprehensible:

- In the **Advanced Options dialog box**, use the arrows to adjust the percentage.

By contrast, to a translator, or to anyone else who is not already familiar with the subject matter, some of the noun phrases in the following sentence might be impossible to understand:

- This catheter is used with an **RF power generator** to deliver RF energy for **intracardiac ablation** of **accessory atrioventricular conduction pathways**.

Nevertheless, it is not appropriate to change noun phrases that are part of the standard vocabulary for a particular subject area. Doing so would make it appear that you are an outsider who doesn't understand the subject matter. Therefore, as the following flowcharts illustrate, the Global English strategy for dealing with noun phrases includes not only revision, but also definition and explanation.

As you can see, the process is simpler for two-word noun phrases (Figure 3.1), because the revision strategies described in guideline 3.5.2 cannot be applied to two-word noun phrases.

Figure 3.1 Flowchart for Two-Word Noun Phrases

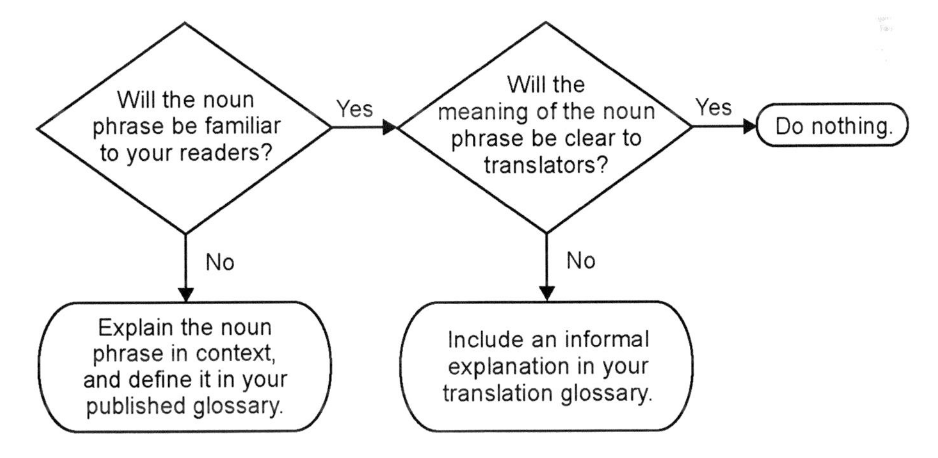

Figure 3.2 Flowchart for Noun Phrases That Are Three or More Words Long

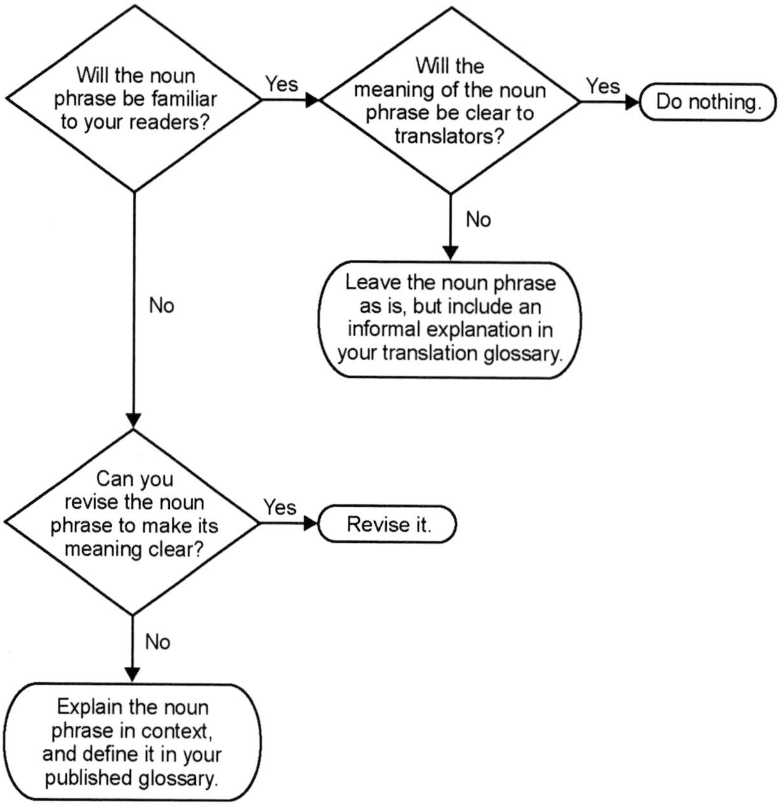

Will the noun phrase be familiar to your readers?
Yes
Will the meaning of the noun phrase be clear to translators?
Yes
Do nothing.
No
Leave the noun phrase as is, but include an informal explanation in your translation glossary.
No
Can you revise the noun phrase to make its meaning clear?
Yes
Revise it.
No
Explain the noun phrase in context, and define it in your published glossary.

Noun Phrases and Human Translation

Human translators can sometimes rely on their knowledge of the everyday world to help them translate noun phrases correctly. For example, a human translator can readily understand that *concrete floor paint* is paint that can be applied to concrete floors rather than *floor paint* that is made of concrete.

But what about a term like *spatial data file*? Is it a data file that is spatial, or is it a file that contains spatial data (that is, data that pertains to spatial coordinates)? The difference in interpretation might seem insignificant in English, but in many languages, the two interpretations are translated differently:

T (French) un fichier de **données spatiales** = a file that contains spatial data

T (French) un **fichier** de données **spatial** = a data file that is spatial

Suppose this term is used in a long document that is translated by a team of translators. If they interpret and translate the term differently, then readers of the translated texts will be confused by the inconsistencies. If the translated document is indexed, then the inconsistencies also make the indexer's job more challenging.

In longer noun phrases, it can be difficult for *anyone* to determine which words are most closely related. Consider the following sentence:

X The **default column pointer location** is column 1.

Does *default column pointer location* mean the location of the default column pointer, or the pointer location of the default column, or the default location of the column pointer? Only a subject-matter expert could know for sure that the last interpretation is correct.

⦸ The **location of the default column pointer** is column 1. [incorrect interpretation]

⦸ The **pointer location of the default column** is column 1. [incorrect interpretation]

✓ The **default location of the column pointer** is column 1.

In many languages, English noun phrases must be translated as a series of prepositional phrases, with one prepositional phrase for each noun or adjective in the noun phrase. The resulting translations can be unwieldy, as the following example illustrates:

X Database Manager Access Descriptor Identification window

T (Spanish) Ventana de Identificación del Descriptor de Aceso de Gestor de Base de Datos

Clearly, it is important to keep noun phrases as short as possible in English. But even the short ones often need to be explained or defined in order for translators to be sure about their meanings. A relatively small amount of effort by the author can save translators a lot of time-consuming research and can help ensure that translations are accurate and consistent.

Noun Phrases and Machine Translation

For machine-translation software, almost every noun phrase that consists of two or more words is potentially ambiguous. In the phrase *hard data*, for example, the adjective *hard* has many possible interpretations, including *not easily penetrated* or *difficult*.

To ensure that appropriate translations are used, most noun phrases must be pre-translated by a human translator and added to the machine-translation software's dictionary. Otherwise, the potential for amusing, embarrassing, or even disastrous mistranslations is very high. For example, Altavista's Babel Fish Translation Web site (http://babelfish.altavista.com/tr) translates *living room furniture* into German as *lebende Raummöbel* (room furniture that is alive).

Noun Phrases and Readability

Long noun phrases pose as much of a problem for native speakers as for translators and machine-translation software. Authors often assume that their primary audience will understand noun phrases like the following. That assumption is rarely correct.

> ✗ The rdfds parameter specifies a **read file descriptor bit mask** that is modified by the call to the Select class.

In many cases, applying one or more simple revision strategies makes a noun phrase much more readable:

> ✓ The rdfds parameter specifies a **bit mask for a read-file descriptor**. The bit mask is modified by the call to the Select class.

In the next example, many readers initially misinterpret *program check handler* as if it were a type of *check handler*:

> ✗ If the subroutine detects an error, it should first pass control to the **program check handler**.

By adding a hyphen, you can make it clear that the noun phrase refers to something that handles *program checks*:

> ✓ If the subroutine detects an error, it should first pass control to the **program-check handler**.

The term *program check* should be defined or explained in a glossary, as discussed in the next section.

3.7.1 Consider defining or explaining noun phrases

As noted earlier, if a noun phrase is part of the standard vocabulary for a particular subject area, then you probably should not change it. Instead, include the noun phrase in one or both of the following glossaries:

- a *published glossary* for your document or product. The structure and location of this glossary depend on the type of document (hard copy, online Help, Web deliverable, and so on).
- a *translation glossary* (for documents that will be translated). Discuss the structure and location of this glossary with your localization company or with your in-house localization coordinator. Translation glossaries are not published; they are for internal use only.

As Figure 3.3 illustrates, the published glossary is a subset of the translation glossary. It should include terms that your primary audience is likely to be unfamiliar with. You might also want to include terms whose definitions are useful for an internal audience, such as other technical writers, editors, and course developers who might be working on the same or related projects. These terms and definitions can also help new employees gain an understanding of the subject matter more quickly.

The translation glossary should include all of the terms that are in your published glossary, plus any additional terms that translators are not likely to be familiar with.

Figure 3.3 Separate Glossaries for Publication and for Translation

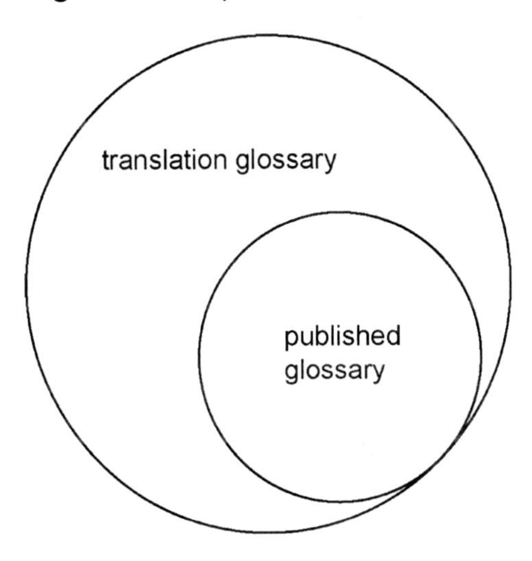

If you are an author, you probably already know your audience and subject area well enough to decide which terms to include in a published glossary. If you are an editor, or if you are an author (such as a technical writer) who has just been assigned to a new subject area, then making such decisions can be more of a challenge. Consult your subject-matter experts or use whatever information sources are available to you to get a better understanding of your subject area and audience. If you are not sure whether to include a term in one or both of these glossaries, then include it.

Note: In order to be of any value to translators, glossaries must be developed and provided *before localization begins.*

What to Include in a Translation Glossary

Translators don't require detailed, formal glossary definitions. An informal explanation, along with the context in which a term is used, is sufficient. Here is an example of an explanation that is suitable for translators:

- contained environment
  an environment in which an applet is given access to only a minimal set of operating-system resources. **Context:** *Java applets run in a contained environment called the Java Sandbox.*

Notice that the explanation contains several technical terms: *environment*, *applet*, *operating system*, and *resources*. But these are all basic software terms that a professional translator will likely be familiar with, or for which the translator can readily find suitable translations in bilingual computer-science glossaries. These terms would not be included in a translation glossary unless they were also part of the published glossary. For example, if the primary audience for the document would not be familiar with the term *applet*, then that term would be defined in the published glossary and would therefore be available to translators.

For industry-specific terms that your primary audience is familiar with, but which translators might not have encountered before, you can glean definitions from the Web or from other published sources. Because you won't be publishing these definitions, you are not infringing on anyone's copyright. However, be sure to record the source of these definitions so that they won't inadvertently be included in any *other* documents that your organization publishes. Here is an example:

- supply chain
  the flow of materials, information, and finances as they move in a process from supplier to manufacturer to wholesaler to retailer to consumer. **Context:** *Supply-chain activities transform raw materials and components into a finished product.* **Source:** http://www.balancedscorecard.biz/Glossary.html

When a noun phrase is three words long or longer, translators need to know which words in the noun phrase are more closely related. For example, is a *graphic device driver* a device driver that is graphic, or is it a driver for a graphic device? The following explanation would be adequate for translators:

▸ graphic device driver
  a driver that controls a graphical output device such as a printer.

The term *driver* can be used in the explanation because it is another basic software term that translators will already be familiar with, or for which they can easily find a definition and a translation.

For translators as well as for primary audiences, definitions or explanations can include any of the following types of terms:

- non-technical terms
- technical terms that are already familiar to the audience
- technical terms that the audience is not familiar with, but which are defined elsewhere in the glossary

3.7.2 Consider revising noun phrases

If a particular noun phrase is not part of the standard vocabulary for your subject area, then you can use one or more of the following revision strategies to simplify and clarify the noun phrase.

Reduction

Before trying to apply any of the other strategies, consider whether you could reduce the length of the noun phrase by eliminating unnecessary words. For example, if an entire document is about SAS/ABC software, it is not necessary to include *SAS/ABC* in the following heading:

✗ Editing **SAS/ABC Table Properties**

✓ Editing **Table Properties**

In the following sentence, unless *MDDB Report Viewer* and *Application Dispatcher* are trademarks (which should be used only as adjectives), the nouns *application* and *software* can be omitted:

✗ The **MDDB Report Viewer application** is a component of **Application Dispatcher software**.

✓ The **MDDB Report Viewer** is a component of the **Application Dispatcher**.

In the next example, it is obvious that the table contains data. Thus, the word *data* can be eliminated from the noun phrase.

✗ The following statement changes the background color of the **LaGuardia table data cells** to yellow.

✓ The following statement changes the background color of the **LaGuardia table cells** to yellow.

Hyphenation

The hyphen is possibly the most under-used punctuation mark in the English language. Authors often assume that readers can use context or knowledge of the natural world to help them interpret noun phrases without any need for hyphens or other syntactic cues.[5] That assumption is often incorrect. Consider this example:

✗ With the aid of a free map, dead celebrity seekers can now tour the grounds of Kenisco Cemetery in Valhalla, N.Y. Highlights include the graves of Lou Gehrig, Danny Kaye, and Ayn Rand.

Whenever an unfamiliar noun phrase begins with an adjective, readers tend to interpret the adjective as modifying the head noun. Thus, in the phrase *dead celebrity seekers*, most readers initially interpret the adjective *dead* as modifying *seekers*, not *celebrity*. In other words, they interpret *dead celebrity seekers* as *celebrity seekers who are dead.*

Knowledge of the natural world should tell readers that this interpretation is incorrect. But research has shown that contextual knowledge doesn't come into play soon enough to prevent readers from misinterpreting sentences or parts of sentences.[6] When readers have to backtrack and reread in order to understand part or all of a sentence, they are left with a negative impression: They feel that the text is confusing even though they were eventually able to figure out the meaning.

In the following revision, a hyphen prevents misreading, making it clear that the celebrities—not the celebrity seekers—are dead:

✓ With the aid of a free map, **dead-celebrity seekers** can now tour the grounds of Kenisco Cemetery in Valhalla, N.Y. High points include the graves of Lou Gehrig, Danny Kaye, and Ayn Rand.

---

[5] Syntactic cues are discussed in more detail in Chapters 6–8 and in Appendix D.

[6] For example, see Ferreira and Clifton (1986) and Rayner, Carlson, and Frazier (1983). This research is summarized in Appendix D, "Improving Translatability and Readability with Syntactic Cues."

Here is a more technical example in which a hyphen makes the noun phrase more readable and easier to translate:

✗ You might want to use a **third party application** such as WinZip to archive the project.

✓ You might want to use a **third-party application** such as WinZip to archive the project.

Rearrangement

Rearrangement is another revision strategy that you can use to clarify noun phrases. In rearrangement, one or more of the words in the noun phrase are placed either in a prepositional phrase or in a relative clause.

In the following example, it is not clear whether the adjective *new* modifies *Knowledge Base* or *location*:

✗ These links point to the **<u>new</u> Knowledge Base location**.

Both of the following revisions eliminate the ambiguity by putting part of the noun phrase into a separate prepositional phrase that begins with *of*. However, only the second interpretation is correct.

⊘ These links point to the **location of the <u>new</u> Knowledge Base**. [incorrect interpretation]

✓ These links point to the **<u>new</u> <u>location</u> of the Knowledge Base**.

Similarly, in the next example, the adjective *recent* modifies the noun *changes*. In the revision, part of the noun phrase is placed in a prepositional phrase that begins with *in*.

✗ The **<u>recent</u> tax law <u>changes</u>** have created confusion.

✓ The **<u>recent</u> <u>changes</u> in tax laws** have created confusion.

If the next example, *European* modifies *customer*, not *data*. Therefore, *data* moves to the beginning, and *European customers* is placed in the prepositional phrase.

✗ The Retail.Europe table contains **<u>European</u> <u>customer</u> data**.

✓ The Retail.Europe table contains **data about <u>European</u> <u>customers</u>**.

Now let's look at an example in which part of a noun phrase is placed in a relative clause (beginning with *that*) rather than in a prepositional phrase. This type of rearrangement is often appropriate when a noun phrase consists entirely of nouns.

✗ CSSTURL contains the URL of the **toolbar frame style sheet**.

✓ CSSTURL contains the URL of the **style sheet that is used for the toolbar frame**.

Using 's to show possession

Using 's is a perfectly suitable way of clarifying some noun phrases. Notice how much easier it is to interpret the following noun phrase after the *'s* has been added:

✗ The **WINDOW field data area** was not defined because not enough memory was available.

✓ The **WINDOW field's data area** was not defined because not enough memory was available.

Disregard the old-fashioned prohibition against using *'s* with nouns that denote inanimate objects. Also disregard the myth that translators or machine-translation software cannot distinguish between possessive *'s* and the use of *'s* as a contraction of *is* or *has*. In formal writing, *'s* is never used as a contraction with nouns.

Substantial revision

Often the best strategy is to substantially revise a sentence that contains a long noun phrase, as in these examples:

✗ The STYLE option specifies the style element to use for **classification variable name headings**.

✓ The STYLE option specifies the style element to use when **names of classification variables** are used as **headings**.

✗ The HBAR statement invokes the **HBAR statistic column label option** and assigns a label to the statistic.

✓ You can use the HBAR statement to assign a **label** to the **column** that contains the **HBAR statistic**.

3.7.3 Always revise noun phrases that contain embedded modifiers

Noun phrases that contain embedded modifiers such as *specific*, *related*, *dependent*, or *oriented* can be especially difficult to translate. Some non-native speakers also find such noun phrases confusing. In most other contexts, these modifiers modify a single noun.

But in the following example, *specific* modifies *data platform*, not just *platform*:

✗ The input engines combine the experimental information with some **data platform specific variables** in the output data set.

Hyphens make the noun phrase more readable, but the hyphens don't make the phrase any easier to translate:

✗ The input engines combine the experimental information with some **data-platform-specific variables** in the output data set.

Instead of using hyphens, restructure your information so that the modifier is no longer embedded in a long noun phrase:

✓ The input engines combine the experimental information with **other variables** in the output data set. **Those other variables are platform specific**.

Sometimes you can eliminate non-essential content and add clarity during the revision process:

✗ For details about data source implementation for each data type such as **data source dependent attributes**, see the appropriate data source reference for data types.

✓ For details about how these data types are implemented for a particular data source, see the documentation for that data source.

Sometimes you can simply eliminate the modifiers with no loss of meaning:

✗ The buckets are used for cash-flow shredding and for other **cash flow related analyses**.

✓ The buckets are used for cash-flow shredding and for other **cash-flow analyses**.

✗ You can add other **data mart related specifications** as needed.

✓ You can add other **data-mart specifications** as needed.

Here are some other examples in which noun phrases that contain embedded modifiers were revised:

✗ Other options, specific to particular file formats, are documented in the **file format specific chapter**.

✓ Options that are specific to a particular file format are documented in the **chapter for that file format**.

✗ Change the attributes for **graph style specific fonts**.

✓ Change the attributes for **fonts that are used in a particular graph style**.

X Filenames must follow **operating system specific conventions**.

✓ Filenames must follow **the conventions of each operating system**.

X Physicians must disclose any financial relationships that they have with companies that manufacture medical devices, implants, pharmaceuticals, or other **medical care related products**.

✓ Physicians must disclose any financial relationships that they have with companies that manufacture any of the following products:

- medical devices
- implants
- pharmaceuticals
- **other products that are used in providing medical care**

Caveats about Revising Noun Phrases

1. Even after using one or more of these strategies, you might still have a noun phrase that needs to be defined or explained in a glossary. For example, in the revision of the following example, even the three-word noun phrase *online access region* is ambiguous. It could mean either a *region of online access* or an *access region that is online*.

    X The only **online access region type** that the software supports is the BMP region.

    ✓ The only **type** of **online access region** that the software supports is the BMP region.

2. A revised noun phrase might not be entirely suitable for other reasons. In the following example, the author was not happy with either of the revisions:

    X **Web-site visitor actions** are recorded in the Web log as a sequence of related events.

    ? The **actions of visitors to Web sites** are recorded in the Web log as a sequence of related events.

    ? The **actions of each visitor to each Web site** are recorded in the Web log as a sequence of related events.

(continued)

The author explained that the phrase *Web-site visitor* was used very frequently in the document that this sentence came from. Using *visitor to a Web site* every time would have been awkward. However, she revised the noun phrase as follows to make it more comprehensible:

✓ The **actions of Web-site visitors** are recorded in the Web log as a sequence of related events.

3.8 Use complete sentences to introduce lists

Priority: HT2, NN3, MT1

Incomplete or interrupted sentences cause extra work for translators, because the order of nouns, verbs, prepositional phrases, and other sentence constituents is different in other languages. In addition, incomplete or interrupted sentences can cause machine-translation software to produce garbled results.

Example 1

X In addition to invoking, managing, and scrolling windows, the windowing environment can

- customize windows
- manage libraries and files
- search text

✓ In addition to invoking, managing, and scrolling windows, the windowing environment can be used as follows:

- to customize windows
- to manage libraries and files
- to search text

Example 2

X In order to use formatting macros, you must be familiar with Base SAS software, including

- using the SAS windowing environment
- creating and submitting simple SAS programs
- using the SAS macro language

✓ In order to use formatting macros, you must be familiar with the following topics:

- using the SAS windowing environment
- creating and submitting simple SAS programs
- using the SAS macro language

For more examples of how to revise lists that violate this guideline, see Appendix C, "Revising Incomplete Introductions to Unordered Lists."

3.9 Avoid interrupting sentences

Priority: HT3, NN3, MT2

Keep subjects and verbs as close together as possible, and avoid any kind of unnecessary interruption to a clause or sentence. These interruptions can separate many different types of sentence constituents, not just subjects and verbs.

3.9.1 Program code, error messages, tables, and figures

In software documentation, technical writers often interrupt sentences with examples of programming code or error messages. In other types of information, authors sometimes interrupt sentences with figures, lists, simple tables, or other types of information that are not part of natural-language sentence structure.

Example 1

X To automatically define a libref each time SAS starts, add

```
libname _saswa <location-of-your-
knowledge-base>;
```

to your autoexec.sas file.

✓ To automatically define a libref each time SAS starts, add the following statement to your autoexec.sas file:

```
libname _saswa <location-of-your-
knowledge-base>;
```

Example 2

X If you get the following error when you try to use Terminal Services, "The specified computer name contains invalid characters. Please verify the name and try again.", then follow the instructions at http://www.cvpn.com/helpdesk/invalid_characters.htm.

✓ If you get the following error when you try to use Terminal Services, then follow the instructions at http://www.cvpn.com/helpdesk/invalid_characters.htm.

```
The specified computer name contains
invalid characters. Please verify the
name and try again.
```

Example 3

X Modify the PROC DISPLAY statement at the beginning of the setup.sas file:

```
proc display c=<libref>.wsvsetup.install.scl;
```

where *libref* is the alias for the new data library.

✓ Modify the PROC DISPLAY statement at the beginning of the setup.sas file:

```
proc display c=<libref>.wsvsetup.install.scl;
```

Replace *libref* with the alias for the new data library.

3.9.2 Adverbs such as *however, therefore,* and *nevertheless*

When they interrupt a clause, conjunctive adverbs such as *however, therefore*, and *nevertheless* interfere with readers' processing. Moreover, many of these adverbs express logical relationships. When a clause is logically related to the preceding clause, tell the reader what the logical connection is at the beginning of the clause.

X You can, **however,** use a RETAIN statement to assign an initial value to any of the previous items.

✓ **However,** you can use a RETAIN statement to assign an initial value to any of the previous items.

3.9.3 Other short sentence interrupters

Here are a few examples of other sentence interrupters that you can usually move to the beginning of a sentence or clause:

for example

✗ You can use a WHERE statement to specify which rows to include in a plot. The following plot**, for example**, displays only rows in which the value of AerobicsClass exceeds 35.

✓ You can use a WHERE statement to specify which rows to include in a plot. **For example**, the following plot displays only rows in which the value of AerobicsClass exceeds 35.

if necessary and *if required*

✗ Manipulate the region of the Linear Fit, **if necessary**, by pressing F3 and selecting `Swap Cursors`.

✓ **If necessary**, manipulate the region of the Linear Fit by pressing F3 and selecting `Swap Cursors`.

Sometimes you must revise a sentence substantially in order to eliminate the interruption:

✗ As each step executes, notes and, **if required,** error messages or warning messages are written to the message area.

✓ As each step executes, notes are written to the message area, **along with any** error messages or warning messages **that are required**.

3.10 Avoid unusual constructions

Priority: HT3, NN2, MT2

The following grammatical constructions are unfamiliar to many non-native speakers of English (including translators), or they are problematic for machine-translation systems, or both. In most cases, it is easy to find alternatives.

3.10.1 The *get* passive

Don't use *get* as an auxiliary verb to form the passive voice. Use a form of the verb *to be* instead.

✗ When you press F6, your program **gets** submitted for execution.

✓ When you press F6, your program **is** submitted for execution.

3.10.2 Causative *have* and *get*

Don't use *have* or *get* as auxiliary verbs meaning *to cause someone to do something* or *to cause something to happen.*

Have

have + a past participle

✗ Highlighting draws attention to values, but you might also want to **have** notes **attached** to those values to explain why the values are highlighted.

✓ Highlighting draws attention to values, but you might also want to **attach** notes to those values to explain why the values are highlighted.

✗ Customers who download the software will **have** the software **activated** during the registration process.

✓ If you download the software, the software **will be activated** when you register it.

✗ All numeric variables that have lengths less than 8 bytes will **have** their lengths **increased** by 1 byte.

✓ The length of any numeric variable that is less than 8 bytes long will **be increased** by 1 byte.

have + an infinitive

✗ **Have** the sponsor **send** a copy of the document to each member of the Project Team.

✓ **Ask** the sponsor **to send** a copy of the document to each member of the Project Team.

Get

get + a past participle

✗ In order to **get** these statistics **printed** in the log, the PRINTALL option must be in effect.

✓ In order for these statistics **to be printed** in the log, the PRINTALL option must be in effect.

✓+ In order to **print** these statistics in the log, **you must specify** the PRINTALL option.

get + an infinitive

✗ Next we tried to **get** the solution engine **to load** vmlinux.

✓ Next we tried **to load** vmlinux on the solution engine.

3.10.3 *In that*

Whenever possible, use *because* instead of *in that*. You might also be able to find other ways to avoid using *in that*.

✗ The order in which you specify the libraries is important, **in that** several libraries define the same symbol.

✓ The order in which you specify the libraries is important **because** several libraries define the same symbol.

3.10.4 *Need not*

Consider the following sentence:

✗ Stored programs **need not** return results.

A non-native speaker might interpret the sentence as *Stored programs need to not return results*, which could be rephrased as *Ensure that stored programs do not return results*. That is an incorrect interpretation, and the *need not* construction can easily be avoided:

✓ Stored programs **do not need to** return results.

3.10.5 Inverted sentences

Except in poetry and questions, it is unusual for the main verb or an auxiliary verb to precede the subject of a clause in English. Although inverting a sentence in this manner can produce a desirable change in emphasis, this construction is confusing to many non-native speakers. It contributes to the problem of unnecessary variation, which Global English strives to minimize, and it can easily be avoided.

✗ Only when stored in an integer variable **is** <u>the value</u> **truncated**.

✓ <u>The value</u> **is truncated** only when it is stored in an integer variable.

In the following example, the sentence must be revised substantially in order to put a subject before the verb:

✗ Implicit in the client/server relationship **is** the network.

✓ The client and the server **are connected** by a network.

3.11 Avoid ambiguous verb constructions

Priority: HT2, NN2, MT2

Some grammatical constructions are ambiguous to machine-translation systems but are readily understood by humans. By contrast, the verb constructions that are discussed in this section are ambiguous and confusing for most readers (both native speakers and non-native speakers), as well as for translators.

3.11.1 *Based on*

In some contexts, *based on* doesn't pose a problem, but in most contexts, it does. It will be easier to understand the problematic contexts if you first look at contexts in which it is not problematic.

Contexts in Which *based on* Is Not Problematic

When *based on* has an easily identifiable subject, it doesn't pose a problem for readers or for translators. In the following example, *based* is an active-voice verb whose subject is *The FCC*:

✓ The FCC **based** the coverage requirement **on** the OET-69 Longley-Rice propagation model.

In the rest of these examples, *based on* is part of a passive verb phrase. But again, the subjects are easy to identify:

✓ This decision will be **based on** an objective assessment of the affected student's neurological development, physical condition, and behavior.

✓ Where possible, these building blocks are **based on** existing ISO standards.

✓ This document is **based on** a review of the currently available data.

Contexts in Which *based on* Is Problematic

- When *based on* is used in an adverbial sense (modifying a verb), its meaning is often unclear. For example, in the following sentence, the *based on* phrase modifies the verb *are merged*. But who can explain what it means?

? These products are merged into a single document, **based on** the original application available on the LEXIS-NEXIS services.

Aside from the problem that this sentence poses for translation, its meaning is impenetrable to everyone except perhaps the author.

As the following examples show, this adverbial usage of *based on* is easy to avoid if you understand what you are really trying to say:

✗ The intersection was built and designed **based on** all the required standards.

✓ The design and construction of the intersection conform to all the required standards.

✗ The products are licensed **based on** your AS/400 serial number and model number.

✓ The product licenses are **based on** your AS/400 serial number and model number.

✗ Sometimes, a diagnosis cannot be made **based on** appearance alone.

✓ Sometimes, appearance alone is not sufficient for making a diagnosis.

- When *based on* occurs at the beginning of a sentence, it is often a dangling modifier. For example, in the following sentence, the *based on* phrase dangles because it modifies the verb in the main clause (*adds*), not the subject (*your employer*):

  ✗ **Based on** your income, your employer adds money to your take-home pay in each paycheck.

  ✓ Your employer adds money to your take-home pay in each paycheck. The amount of money that is added **depends on** your income.

  Again, it is usually easy to find alternatives to this use of *based on*:

  ✗ **Based on** their other qualifications, we might decide to train applicants to take the C.D.L.

  ✓ If applicants are otherwise well qualified, we might decide to train them to take the C.D.L.

  ✗ **Based on** our experiences in Romania in 1994 and 1996, we expected students to vary widely in English proficiency.

  ✓ **Because of** our experiences in Romania in 1994 and 1996, we expected students to vary widely in English proficiency.

- When *based on* follows a noun, it is often unclear whether it modifies that noun or a verb that occurs earlier in the sentence. For example, in the following sentence, *based on* could be modifying either *discipline policy* or *has established*:

✗ The college has established a discipline policy **based on** the rights of others.

If *based on* modifies the noun that it follows, then make it unambiguous by expanding it into a relative clause:

✓ The college has established a discipline policy **that is based on** the rights of others.

Here are some additional examples:

✗ Your feedback helps us develop new products and services **based on** your needs.

✓ Your feedback helps us develop new products and services **that are based on** your needs.

✗ Sexism and racism often compound discrimination **based on** disabilities.

✓ Sexism and racism often compound discrimination **that is based on** disabilities.

Be careful when applying this revision strategy, because *based on* doesn't always modify the noun that it follows. In the following example, a more drastic revision is required:

✗ To perform an action **based on** a condition, use an IF-THEN statement.

⊘ To perform an action **that is based on** a condition, use an IF-THEN statement.

✓ If you want to perform an action **only when a particular condition exists**, use an IF-THEN statement.

The strategy of expanding past participles such as *based* into relative clauses is also discussed in guideline 6.6.1, "Revise past participles that follow and modify nouns."

See guideline 7.3, "Revise dangling -ING phrases," for a more detailed discussion of a similar type of dangling modifier.

3.11.2 *Require* + an infinitive

When a form of the verb *require* is followed by an infinitive, the result can be a type of ambiguity that most authors overlook. Consider the following sentence:

✗ Two technicians are **required to perform** maintenance on fuser assemblies.

This sentence could be interpreted in the following ways:

- **?** Of the entire staff of technicians, two **must perform** maintenance on fuser assemblies. (The other technicians don't have this responsibility.)
- **?** In order to perform maintenance on fuser assemblies, two technicians **are required**. (One technician cannot do the job alone.)

This type of ambiguity confuses many readers. Human translators often cannot determine which interpretation is correct, if indeed they even notice the ambiguity. And the only way to know how a machine-translation system will translate the original sentence is to submit the sentence and then examine the output.

Sometimes you can eliminate the ambiguity by changing *required to* to *required for* plus a gerund, as in this example:

- **X** Data storage space is the amount of space on a disk or tape that is **required to store** data.
- **✓** Data storage space is the amount of space on a disk or tape that is **required for storing** data.[7]

Sometimes changing *required to* to *required in order to* solves the problem:

- **X** Several double-byte character variables are **required to produce** the table.
- **✓** Several double-byte character variables are **required in order to produce** the table.
- **X** The LP option is **required to produce** line-printer output.
- **✓** **In order to produce** line-printer output, you must specify the LP option.

But often the best approach is to use a more substantial revision, as in this example:

- **X** The LOOP statement does not **require** additional statements **to stop** processing records.
- **?** The LOOP statement does not **require that** additional statements **stop** processing records. (probably an incorrect interpretation)
- **✓** When the last record is processed, the LOOP statement automatically **stops** executing. No additional statements **are required**.

---

[7] The *required for* + gerund + noun construction (as in *required for storing data*) can still be ambiguous to machine-translation software, but in most contexts it is clear to human readers and translators.

3.11.3 *Appear* + an infinitive

In technical documentation, the verb *appear* usually means *to become visible*. But when it is followed by an infinitive, it is often understood to mean *to seem*, as in this example:

> X The Message Display window **appears to indicate** how many records were actually inserted into the new table.

If you replace *appears to* with *seems to*, it is obvious that this meaning is not what the author intended:

> ⊘ The Message Display window **seems to indicate** how many records were actually inserted into the new table.

This sentence can easily be revised to avoid the ambiguity:

> ✓ The Message Display window **appears. This window** indicates how many records were actually inserted into the new table.

3.11.4 *Has* or *have* + past participle + noun phrase

When a past participle immediately follows *has* or *have*, it is usually acting as part of the verb, as in this example:

> ▶ Sanjay Patillo **has agreed** to lead the investigation.

But what happens when this combination of *has* or *have* plus a past participle is immediately followed by a noun phrase?[8] Sometimes it is not clear whether the past participle is part of the verb, or whether it is acting as an adjective, as in this sentence:

> X Components **have published interfaces** for communication and standard methods for retrieving information.

Most readers have to back up and reread the sentence because they initially interpret *published* as part of the verb rather than as an adjective that modifies the noun *interfaces*. In other words, they interpret *Components have published interfaces* as the active-voice equivalent of *Interfaces have been published by components*. When they encounter the word *for*, they realize that *published interfaces* actually means *interfaces that are (or that have been) published.*

[8] As explained in "An Overview of Noun Phrases" on page 44, a noun phrase can consist of a single noun.

The sentence should be revised to eliminate the ambiguity:

✓ For each component, DevApp software **provides published interfaces** for communication, as well as standard methods for retrieving information.

Here is another example:

X Each column **has assigned attributes** such as name, type, and length.

In this case, the *has* or *have* + *past participle* + *noun phrase* construction is followed by *such as* . . .. The *such as* construction helps most human readers quickly recognize that the past participle is acting as an adjective. But the sentence is still ambiguous to machine-translation software and to some translators. Thus, it is still a good idea to revise the sentence. Here are two possible revisions:

✓ Attributes such as name, type, and length **are assigned to** each column.

✓ Each column **has** attributes such as name, type, and length.

The *has* or *have* + past participle + noun phrase construction is not ambiguous if the noun phrase starts with an article (*a*, *an*, or *the*) or with some other determiner. Therefore, you can sometimes resolve the ambiguity by inserting a determiner after the participle, as in these examples:

X Scientists **have refined techniques** for measuring ozone depletion.

✓ Scientists **have refined <u>the</u> techniques that they use** for measuring ozone depletion.

X Business accounts and Professional DSL accounts **have assigned IP addresses**.

✓ **Each** Business account and Professional DSL account **has <u>an</u> assigned IP address**.

X I **have extracted stills** from the video to highlight the most important scenes.

✓ I **have extracted <u>several</u> stills** from the video to highlight the most important scenes.

A Similar but Rare Construction

This same type of ambiguity can occur with the verb *to be* + past participle + noun phrase.

> ✗ Participants **are** usually **assigned** counselors on the first day of camp.

However, if you follow guideline 6.8, "Use *to* with indirect objects," then you will revise the sentence as follows, and the ambiguity will disappear:

> ✓ Counselors **are** usually **assigned to** participants on the first day of camp.

3.11.5 *Has* or *have* + noun phrase + past participle

Unlike the similar construction in guideline 3.11.4, this unusual construction is not likely to cause misreading. However, it is unnecessarily complex, and it is easy to avoid. Why not make translators' lives a bit easier by eliminating it?

> ✗ If you **have** access control **enabled**, then you must specify a valid user name and password.

> ✓ If access control **is in effect**, then you must specify a valid user name and password.

> ✗ An ITSV server is a host that **has** ITSV software **installed** with a server license.

> ✓ An ITSV server is a host **on which** ITSV software **has been installed** with a server license.

> ✗ If you **have** a USER library **defined**, you can still use the WORK data library by specifying WORK.*SAS-data-set*.

> ✓ If you **have defined** a USER library, you can still use the WORK data library by specifying WORK.*SAS-data-set*.

> ✗ The software already **has** the printing functions **built in**.

> ✓ The printing functions **are** already **included in** the software.

3.11.6 *Must be, must have,* and *must have been*

The modal verb *must* can convey an obligation (as in *I must be going*) or a logical necessity (as in *I must be dreaming* or *You must have been a beautiful baby* or *You must enjoy doing things the hard way*). The latter interpretation is unusual in scientific and technical writing, but the potential for ambiguity is worth noting. Consider the following sentence:

✗ If you are accessing data that is owned by another user, you **must have been granted** access privileges.

If *must* indicates a logical necessity, then you can convey that meaning by using the word *presumably* instead:

? If you are accessing data that is owned by another user, then **presumably you have been granted** access privileges.

If *must* indicates an obligation, then you can rephrase the sentence as follows:

? In order for you to access data that someone else owns, the owner **must first grant** access privileges to you.

In the next example, the *must have been* ambiguity is eliminated by revising the sentence substantially:

✗ If the brushes have stopped turning, then you **must have broken** the connection between the vacuum cleaner attachment and the canister.

✓ The brushes stop turning, then **check to see whether** the connection between the vacuum cleaner attachment and the canister **has been broken**.

3.12 Write positively

Priority: HT2, NN2, MT2

In cautions and warnings, negative statements are often necessary and appropriate. In other contexts, if you can accurately express an idea either positively or negatively, then express it positively. Positive statements are often more succinct, in addition to being easier to comprehend and translate.

✗ Add a statement to the program to specify that messages **not be displayed**.

✓ Add a statement to the program **to suppress messages**.

Also avoid including more than one *not* in a sentence. As the following sentence illustrates, double-negative constructions can be confusing for non-native speakers, translators, and native speakers alike:

✗ Do **not** make changes to the CODES table, or the changes will **not** be saved when the table is re-created.

The following revision is much clearer:

- ✓ Changes that you make to the CODES table will **not** be saved when the table is re-created.

Putting the two negatives in separate sentences would also be acceptable:

- ✓ Do **not** make changes to the CODES table. Your changes will **not** be saved when the table is re-created.

Here are some other examples in which positive statements (or statements that are less negative) are simpler and more direct:

- ✗ You **cannot** access databases for which you **do not have** appropriate permissions.
- ✓ You **can** access **only** databases for which you **have** appropriate permissions.
- ✗ Applications will **not** be accepted if any information is **not** completed.
- ✓ **Incomplete** applications will **not** be accepted.

Chapter 4

Using Modifiers Clearly and Carefully

Introduction

Unless you once learned to diagram sentences and have managed to retain that knowledge, you probably are not accustomed to thinking about how different parts of a sentence relate to each other. But anyone who wants to master Global English must develop a heightened awareness of these relationships. The guidelines in this chapter will help you develop that awareness.

4.1 Place *only* and *not* immediately before whatever they are modifying

Priority: HT3, NN3, MT1

In general, place the one-word modifiers *only* and *not* immediately in front of whatever they are modifying. Incorrect placement of these modifiers is problematic for the following reasons:

- Human translators don't always notice that these modifiers are in the wrong location. If they don't notice a misplaced modifier, then they produce incorrect translations. In some cases, they cannot be sure what the author meant. Seeking clarification on such issues is very time-consuming.
- Machine-translation software translates these modifiers according to their placement in a clause or sentence. Incorrect placement leads to mistranslations.

- Because some authors use these modifiers correctly and others do not, inconsistent placement of these modifiers causes unnecessary variation and therefore makes translation less efficient. (See "Translation Memory" in Chapter 1, "Introduction to Global English.")
- It's simply bad English!

4.1.1 *Only*

In this example, the author did not mean that this type of forecasting only *works*, as opposed to only doing something else:

> ✗ Artificial Neural Network forecasting **only** works in Version 7.0.

She meant that it works only in Version 7.0, not in some other version. In other words, *only* is modifying the prepositional phrase *in Version 7.0* and should therefore be placed immediately in front of it:

> ✓ Artificial Neural Network forecasting works **only** in Version 7.0.

In this next example, *only* is modifying the *during* phrase, not *in effect*:

> ✗ Syntax-checking mode is **only** in effect during the step in which the program encountered the error.

> ✓ Syntax-checking mode is in effect **only** during the step in which the program encountered the error.

And in this example, *only* modifies the *within* phrase, not *effective*:

> ✗ Treatment for the most common type of stroke is generally **only** effective within three hours of the first symptoms.

> ✓ Treatment for the most common type of stroke is generally effective **only** within three hours of the first symptoms.

⚠ An Exception to This Guideline

There is one context in which a violation of guideline 4.1.1 is idiomatic. The context can be described as follows:

subject + the verb *to be* + prepositional phrase beginning with *for* + *only*

Here is an example:

▸ The number below is for credit card orders **only**.

Because *only* follows the prepositional phrase that it modifies, this sentence violates guideline 4.1.1. The correct form of the sentence would be as follows:

✓ The number below is **only** for credit card orders.

However, because the pattern of the original sentence is so common and idiomatic, you might choose to allow the apparent misplacement of *only* in that context.

When a preposition other than *for* is used in this sentence pattern, placing *only* in front of the prepositional phrase is perfectly natural and conforms to guideline 4.1.1:

✗ The logo is on the front **only**.

✓ The logo is **only** on the front.

✗ Sorry, this Web site is in English **only**.

✓ Sorry, this Web site is **only** in English.

4.1.2 *Not*

Not is another one-word modifier that should be placed immediately in front of whatever it is modifying. Consider the following example:

✗ All information requests have **not** crossed my desk, but quite a few have.

Saying that *all information requests have not crossed my desk* is equivalent to saying that none of them has crossed your desk. Here's what the author meant:

✓ **Not** all information requests have crossed my desk, but quite a few have.

Similarly, saying that *everyone is not getting rich* is the same as saying that no one is getting rich:

✗ The fact is that everyone is **not** getting rich. On the contrary, the world is full of disappointed investors.

The word *not* should be placed before *everyone*, because that's what it modifies:

✓ The fact is that **not** everyone is getting rich. On the contrary, the world is full of disappointed investors.

This incorrect placement of *not* is just as illogical as the omission of *not* in the sentence *I could care less*. Many native speakers of English use that sentence to mean exactly the opposite: *I could not care less*. Obviously, saying the opposite of what you intended to say is not good Global English!

4.2 Clarify what each prepositional phrase is modifying

Priority: HT2, NN2, MT2

In guideline 4.1 you saw that you could correct modification problems by placing one-word modifiers in front of what they are modifying. By contrast, prepositional phrases usually *follow* the words that they modify.

Unfortunately, prepositional phrases don't have to *immediately* follow whatever they are modifying. As a result, they are a source of ambiguity that often confuses readers and that can lead to incorrect translations.

Consider the following example:

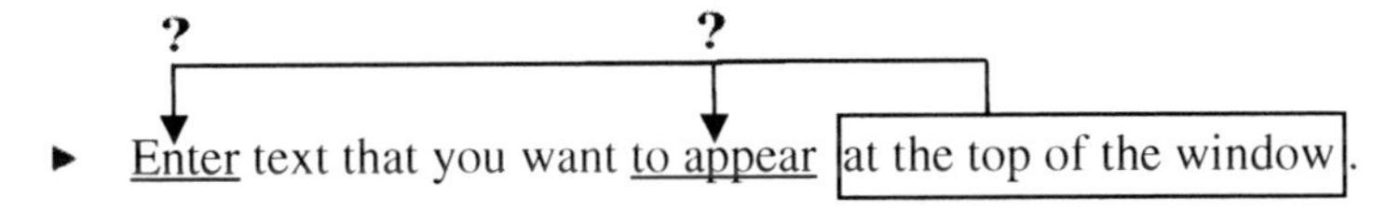

Do the prepositional phrases (*at the top of the window*) modify *enter* or *to appear*? That is, are they describing the location of a box or field in which the user is supposed to enter some text? Or are they describing where that text will be displayed?

If the above sentence is used in a software user interface, then it might be one of hundreds of sentences and phrases in a Java `resource.properties` file (or in some other type of file). Translators often have to translate these texts without being able to see the interface. In other words, translators sometimes have little or no context to draw from. A translator might translate an ambiguous sentence incorrectly without even noticing the ambiguity.

To machine-translation software, even the following sentence is ambiguous, because *on the View menu* could be modifying either *data* or *Put*:

► Put commands that change the user's view of data **on the View menu**.

Guidelines 4.2.1–4.2.5 constitute a procedure that you can follow in order to identify and resolve such ambiguities. Figure 4.2 gives you an overview of the procedure.

Notice that asking yourself whether the context makes the modification clear is the fourth step in the procedure, not the first. A sentence that is clear and readable to you might not be clear and readable to readers who are less familiar with the subject matter. Therefore, you should always go through the first three steps of the procedure.

Figure 4.2a Flowchart for Clarifying Prepositional Phrase Modification

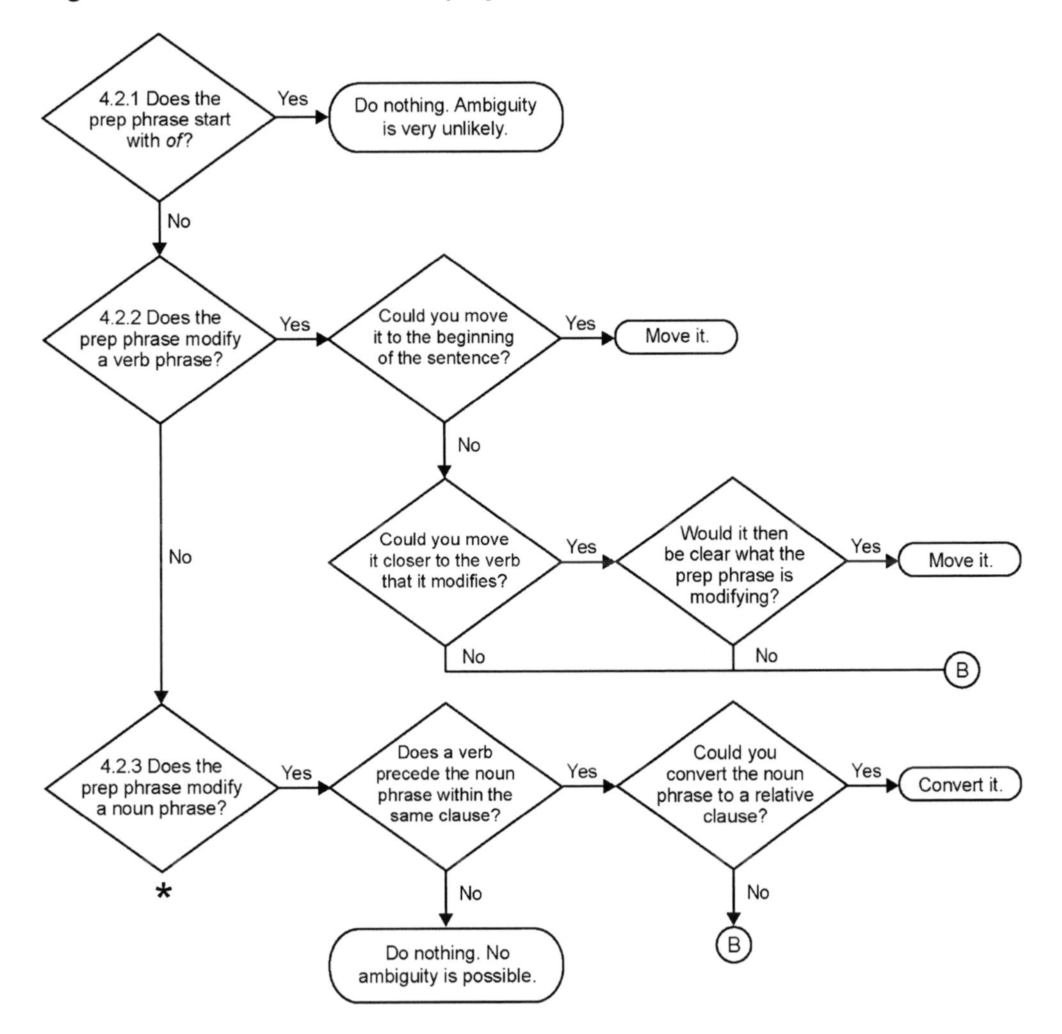

* A No response is not possible here. If a prep phrase does not modify a verb phrase, the only other possibility is that it modifies a noun phrase.

Figure 4.2b Flowchart for Clarifying Prepositional Phrase Modification

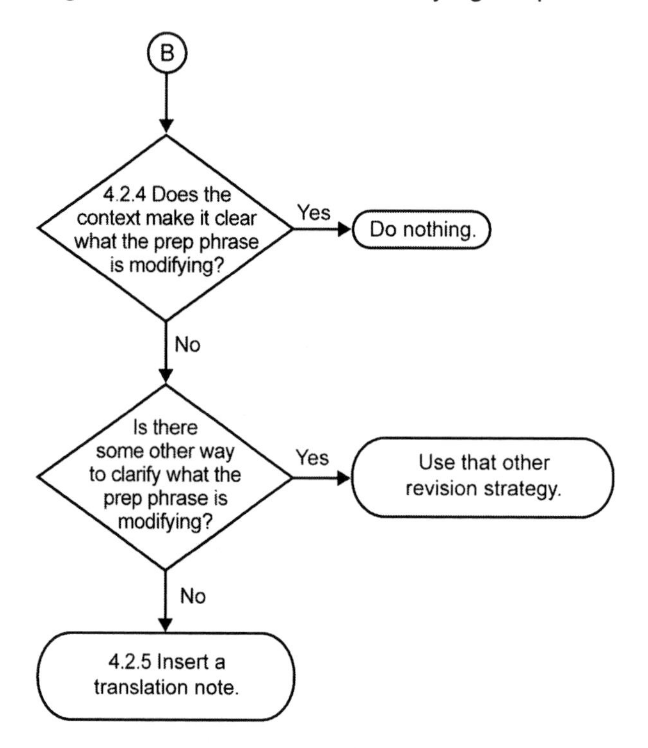

4.2.1 If the prepositional phrase starts with *of*, then do nothing

Prepositional phrases that start with *of* are unambiguous. In technical documentation, they virtually always modify noun phrases, and they always immediately follow whatever they are modifying:

- The RANGE= option displays the range **of numeric data** that each bar represents on the axis **of the chart**.
- Put commands that change the user's view **of data** on the View menu.
- For a text version **of this Web site**, go to text.uchicago.edu.

4.2.2 If the prepositional phrase modifies a verb phrase, consider moving it

Move the prepositional phrase to the beginning of the sentence

A prepositional phrase that modifies a verb phrase is potentially ambiguous if there is an intervening noun phrase. For example, in the following sentence, *on a standard tape label* could be modifying either the verb phrase *are available* or the intervening noun phrase *table name*:

✗ Only 17 characters are available for the table name **on a standard tape label**.

In other words, does *on a standard tape label* describe where the table name is located, or does it describe where the 17 characters are available?

When you move the prepositional phrase to the beginning of the sentence, the only possible interpretation is that it modifies *are available*:

✓ **On a standard tape label,** only 17 characters are available for the table name.

Move the prepositional phrase closer to the verb that it modifies

If a noun phrase separates a prepositional phrase from the verb that the prepositional phrase is modifying, then try to put the prepositional phrase closer to the verb. In the following sentence, *with other users* modifies *to share*, but it could be interpreted as modifying the intervening noun phrase *platforms*:

✗ The server enables a client user to share data across platforms **with other users**.

The second interpretation would be synonymous with the following revision:

⊘ The server enables a client user to share data across platforms **that have other users**.

You can prevent readers and translators from misinterpreting the sentence by moving *with other users* closer to *to share*:

✓ The server enables a client user to share data **with other users** across platforms.

This revision strategy doesn't always work, so use your best judgment about when to apply it. For example, in the following sentence, *with new data* modifies the infinitive verb *to update*:

▸ You can use the DATA step to update an existing table **with new data**.

When you place the prepositional phrase immediately after the infinitive, the revision sounds very unnatural:

⊘ You can use the DATA step to update **with new data** an existing table.

As noted in "The Cardinal Rule of Global English" in Chapter 1, if you are not sure whether a revision sounds natural, consult colleagues who are native speakers of English.

4.2.3 If a prepositional phrase modifies a noun phrase, consider expanding it into a relative clause

A prepositional phrase that modifies a noun phrase is unambiguous if no verb occurs earlier in the same clause. For example, in the following sentence, *in the main menus* is unambiguous because there is no preceding verb that it could be modifying. The only thing it can be modifying is the noun *functions*:

▶ Many of the functions **in the main menus** are also available in pop-up menus.

By contrast, the prepositional phrase in the next sentence is preceded both by a noun (*data*) and a verb (*prints*):

✗ The PRINT statement prints the data **in the SPSS file**.

The sentence is initially confusing to human readers and translators, and it is certainly ambiguous to machine-translation software, because *in the SPSS file* could be modifying either *data* or *prints*. That is, the prepositional phrase could be describing either the location of the data or the location at which the printing occurs.

To resolve the ambiguity, you can convert the prepositional phrase to a restrictive relative clause[1] by inserting *that is*:

✓ The PRINT statement prints the data **that is in the SPSS file**.

Restrictive relative clauses are less problematic than prepositional phrases because they modify only noun phrases.

As always, consider whether some other type of revision would work better:

✓+ The PRINT statement prints the contents **of the SPSS file**.

[1] See guideline 4.3, "Clarify what each relative clause is modifying," for an explanation of restrictive relative clauses.

Sometimes you need to include an active or passive verb in the relative clause:

✗ A U.S. appeals court ruled that makers of file-sharing software cannot be held liable for copyright infringement **by their users**.

✓ A U.S. appeals court ruled that makers of file-sharing software cannot be held liable for copyright infringements **that their users commit**.

✗ When you access a function **in an external DLL**, you transfer control to the external function.

✓ When you access a function **that is stored in an external DLL**, you transfer control to the external function.

If the prepositional phrase begins with *with* and modifies a noun, then often you can change it to a relative clause by using *that has*, *that have*, *that contains*, or *that contain*:

✗ The following code dynamically creates an HTML table **with** four cells.

✓ The following code dynamically creates an HTML table **that has** four cells.

✗ Use a LENGTH statement when you have a large data set **with** many character variables.

✓ Use a LENGTH statement when you have a large data set **that contains** many character variables.

⚠ Careful Analysis Is Required

Before converting any prepositional phrase to a relative clause, make sure that the prepositional phrase modifies the noun phrase that you think it modifies. Otherwise you might distort the meaning of the sentence that you are revising.

For example, in the following sentence, the prepositional phrase (*with new data*) modifies the verb *update*, not the noun *table*.

✓ You can use the DATA step to update an existing table **with** new data.

Therefore, the following revision would be incorrect:

⦸ You can use the DATA step to update an existing table **that has** new data.

Because *update* implies *with new data*, the best solution would be to omit the prepositional phrase. The word *existing* is also unnecessary:

✓+ You can use the DATA step to update a table.

4.2.4 If readers and translators can determine what the prepositional phrase is modifying, then do nothing

After following the strategies that are explained in 4.2.1–4.2.3, consider whether a typical reader or human translator will be able to correctly identify what the prepositional phrase is modifying. If so, then you have solved the modification problem adequately.

If you are using machine-translation software, it's unlikely that the software will always correctly identify what prepositional phrases are modifying no matter what you do. Therefore, the goal here is merely to make the modification clear enough that a human translator could make any necessary corrections in the machine-translation output.

4.2.5 When necessary, insert a translation note

Sometimes there is no elegant way of making a sentence unambiguous. For example, in the following sentence, *in the navigation tree* could be modifying either *an object* or *select*:

✗ If you select an object **in the navigation tree** that contains subfolders, the display area lists those subfolders.

Suppose that it modifies *an object*. If you follow guideline 4.2.3, the revision will read as follows:

⊘ If you select an object **that is in the navigation tree** *that contains subfolders*, the display area lists those subfolders.

The revised sentence is stylistically unacceptable. In addition, the revision now leads readers to interpret *that contains subfolders* as modifying *navigation tree* rather than *object*. That interpretation is incorrect. You could easily spend twenty minutes trying to find an acceptable revision (possibly without success), and you probably don't always have time to strive for perfection.

In this case, inserting a translation note[2] is probably the best solution:

✓ If you select an object in the navigation tree that contains subfolders, the display area lists those subfolders.`<translationNote>"in the navigation tree" modifies "object," not "select."</translationNote>`

[2] For more information about translation notes, see "Insert Explanations for Translators" in Chapter 1.

4.3 Clarify what each relative clause is modifying

Priority: HT2, NN2, MT2

This guideline pertains to *restrictive relative clauses* like the following:

- Butterfly taxa **that have a low risk of extinction** are listed in the Least Vulnerable column.

By contrast, a *non-restrictive relative clause* begins with the relative pronoun *which*, *who*, *whom*, or *whose* and is preceded by a comma:

- Butterfly taxa, **which** have a low risk of extinction, are listed in the Least Vulnerable column.

Guideline 5.3, "Don't use *which* to refer to an entire clause," pertains to this non-restrictive use of *which*.

Unlike prepositional phrases, restrictive relative clauses (referred to in the rest of this section simply as relative clauses) always modify noun phrases. However, a relative clause doesn't always immediately follow the noun phrase that it modifies. For example, if two relative clauses are joined by a coordinating conjunction (*and*, *or*, or *but*), then the second one doesn't immediately follow the noun phrase that it modifies:

- This article describes features **that facilitate collaboration** but **that are not intended to increase security**.

However, one context in which a relative clause is likely to be ambiguous or confusing is when it follows one or more prepositional phrases. For example, the following sentence is initially confusing to many readers. The relative clause could be modifying either the closest preceding noun phrase (*products*) or a noun phrase that occurs earlier in the sentence (*descriptions*):

✗ Click `Overview` to view short descriptions *of the* products **that include information links**.

✓ Click `Overview` to view short descriptions of the products. Each description includes links to more-detailed information.

A relative clause can also be ambiguous when it follows a series, as in this example:

✗ The audience consisted of *employees, quality partners, and customers* **who are participating in beta testing**.

Does the relative clause modify only *customers*, or does it also modify *employees* and *quality partners*? That is, are all three groups participating in beta testing? If only customers are participating, then you can eliminate the ambiguity by using an unordered list:

✓ The audience consisted of three groups of people:

- employees
- quality partners
- customers who are participating in beta testing

If all three groups are participating in beta testing, then you must restructure the sentence in order to make it unambiguous:

✓ The audience consisted of three groups of people who are participating in beta testing: employees, quality partners, and customers.

In some cases, the verb in the relative clause agrees in number with the noun phrase that the relative clause modifies. As a result, the following sentence is unambiguous. The verb *are* in the relative clause can refer only to the plural noun phrase, *lines*, not to the singular noun phrase, *statement*:

✓ The Edit window accepts lines *in a DATALINES statement* **that are longer than 256 characters**.

But there is another type of relative clause in which the relative pronoun *that* is the object of the verb, not the subject. In that case, even if the relative clause has a singular verb, the relative clause is not necessarily modifying a singular noun phrase. For example, in the following sentence, *someone else* is the subject of the relative clause:

✗ If you change chapters of a document **that someone else has checked out**, you won't be able to save your changes.

The verb in the relative clause, *has checked out*, has to agree with *someone else*, not with the noun that the relative clause is modifying. Thus, the relative clause is ambiguous because it could be modifying either *chapters* or *document*. Here are two possible revisions:

✓ If you change chapters **that someone else has checked out**, you won't be able to save your changes.

✓ If someone has checked out a document, then you cannot save any changes that you make to any chapters of that document.

Related Guideline

- 6.5, "Clarify which parts of a sentence are being joined by *and* or *or*"

4.4 Use *that* in restrictive relative clauses

Priority: HT2, NN3, MT2

For consistency, always use *that*, not *which*, with restrictive relative clauses:

✗ A DBMS client is an application **which** manages connections to a specific DBMS.

✓ A DBMS client is an application **that** manages connections to a specific DBMS.

✗ The following table lists the types of graph objects **which** SAS/STAT procedures produce.

✓ The following table lists the types of graph objects **that** SAS/STAT procedures produce.

Also consider making other revisions to ensure clarity and improve style:

✗ The Name field displays the name of the component as specified in the Composite Definition **which** should receive the method.

✓ The Name field displays the name of the component (as specified in the Composite Definition) **that** should receive the method.

✗ After you specify a desirability function for each response, ADX develops a single composite response **which** is the geometric mean of the desirabilities of the individual responses.

✓ After you specify a desirability function for each response, ADX develops a single composite response **that** is the geometric mean of the desirabilities of the individual responses.

✓+ After you specify a desirability function for each response, ADX develops a single composite response. **That composite response** is the geometric mean of the desirabilities of the individual responses.

4.5 Consider moving anything that modifies a verb to the beginning of the clause or sentence

Priority: HT3, NN3, MT2

In 4.2.2 you saw that moving a prepositional phrase to the beginning of a sentence made it clear that the prepositional phrase was modifying the verb in the sentence rather than a noun phrase. The same is true of other types of modifiers that are described in the following sections.

4.5.1 Participial phrases

In the sentence below, the participial phrase describes how a software user should write a SYMBOL statement. In other words, the participial phrase modifies the verb *Write* and can therefore be moved to the beginning of the sentence:

✗ Write a SYMBOL statement, **using the fewest options possible**, that changes the color of the plotting symbols to RED.

✓ **Using the fewest options possible**, write a SYMBOL statement that changes the color of the plotting symbols to RED.

4.5.2 *In order to*

You can sometimes move *in order to* phrases to the beginning of a sentence as well:

✗ A server must have significant computing resources **in order to process large tables rapidly**.

✓ **In order to process large tables rapidly,** a server must have significant computing resources.

4.5.3 Adverbial phrases

You can also relocate various types of adverbial phrases, as in this example:

✗ Avoid creating situations that require a message box to be displayed **whenever possible**.

✓ **Whenever possible,** avoid creating situations that require a message box to be displayed.

4.6 Clarify ambiguous modification in conjoined noun phrases

Priority: HT2, NN3, MT2

Conjoined noun phrases (noun phrases that are joined by the coordinating conjunctions *and* or *or*) sometimes contain a type of ambiguity that often goes unnoticed. For example, in the following sentence, hardly anyone questions whether *apple* modifies *ice cream*, or whether it modifies only *pie*:

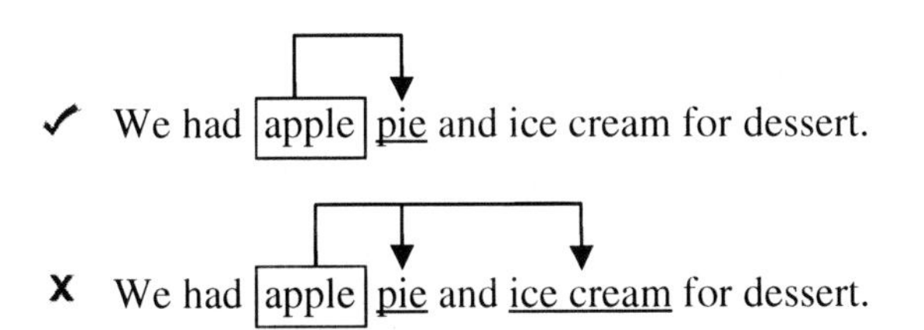

Because *apple pie* is a common compound noun, and because apple-flavored ice cream is a rarity (at least in the United States), most readers don't even consider the possibility that *apple* could be modifying *ice cream.*

But the less you know about a topic, the less certain you can be about the relationship between the word or words that precede *and* or *or* and the noun phrase or noun phrases that follow. For example, consider the following sentence:

▸ Lavoie plans to keep all the backyard plantings, including **fragrant spearmint** and **ginkgo biloba trees.**

Either of the following interpretations would be possible, but neither is correct:

? Lavoie plans to keep all the backyard plantings, including **fragrant spearmint** trees and **ginkgo biloba trees.**

? Lavoie plans to keep all the backyard plantings, including **fragrant spearmint** trees and fragrant **ginkgo biloba trees.**

An easy way to make it clear that spearmint is not a tree and that ginkgo biloba trees are not fragrant is to insert *the* before each noun phrase:

✓ Lavoie plans to keep all the backyard plantings, including the fragrant spearmint **and** the ginkgo biloba trees.

This next sentence might be clear to human readers and translators, but it would be problematic for machine-translation software because *charges* can be either a noun or a verb:

> X The statement shows your credits **and** charges for the month to date.

The software might interpret the sentence as follows:

> ⊘ The statement shows your credits, **and** [the statement] charges for the month to date.

By repeating *your*, you can make the sentence unambiguous even for machine-translation software:

> ✓ The statement shows your credits **and** your charges for the month to date.

Sections 4.6.1–4.6.7 provide more-detailed explanations of these and other ways of clarifying ambiguous modification in conjoined noun phrases.

Remember that the cardinal rule of Global English always takes precedence, and don't apply any of the revision strategies until you understand all of them.

The Cardinal Rule of Global English

Don't make any change that will sound unnatural to native speakers of English.

Corollary

There is almost always a natural-sounding alternative if you are creative enough (and have enough time) to find it!

4.6.1 Consider using identical grammatical structures in each noun phrase

If a modifier modifies both or all of the conjoined noun phrases, then consider using the same grammatical structure in each noun phrase. For example, in the following sentence, the modifier *average* modifies both *ages* and *salaries*:

> X Table 8.1 shows **average ages** and **salaries** for employees in each division.

To make the modification unambiguous, repeat the modifier so that you have a modifier plus a noun phrase on both sides of the conjunction:

> ✓ Table 8.1 shows **average ages** and average **salaries** for employees in each division.

In some contexts, you must repeat an article as well as a modifier in order to make the grammatical structures identical. In the next example, the article *the* and the modifier *right* are both repeated:

- X Now that we have **the right people** and **strategy**, we know that we will be successful.
- ✓ Now that we have **the right people** and the right **strategy**, we know that we will be successful.

In other cases, you must repeat a head noun:

- X The data typically includes either **HTML or XML elements**.
- ✓ The data typically includes either **HTML** elements **or XML elements**.
- X **Variable**, **fixed**, and **weighted values** cannot be changed in the interface.
- ✓ **Variable** values, **fixed** values, and **weighted values** cannot be changed in the interface.

In some contexts, you must also change the head noun from plural to singular:

- X You can divide **the Resource module**, **Activity module**, and **Cost Object module pages** into up to three panes.
- ✓ You can divide **the Resource module** page, the **Activity module** page, and the **Cost Object module** page into up to three panes.

If the first noun phrase includes an article, then you probably need to include articles in the subsequent noun phrases as well:[3]

- X **The CPU** and **memory usage** for the two programs are about the same.
- ✓ **The CPU** usage and the **memory usage** for the two programs are about the same.

---

[3] This statement is probably always true, but it has not yet been verified through further data collection and analysis.

✗ Before changing **a shared**, **unique**, or **driver quantity**, copy each quantity to a destination account.

✓ Before changing **a shared** quantity, a **unique** quantity, or a **driver quantity**, copy each quantity to a destination account.

4.6.2 Consider inserting an article after the conjunction

In the examples in 4.6.1, the modifier modified both or all of the conjoined noun phrases. If a modifier modifies the first noun phrase, but not a subsequent noun phrase, then you must use a different revision strategy. For example, in the following sentence, the adjective *logical* modifies only *member name*, not *member type*:

▸ Use the FILENAME statement to specify **the logical member name** and **member type**.

Therefore, the 4.6.1 revision strategy would be incorrect:

⊘ Use the FILENAME statement to specify the **logical member name** and the logical **member type**.

Instead, you can insert *the* following *and* to prevent *logical* from modifying the second noun phrase:

✓ Use the FILENAME statement to specify **the logical member name** and the **member type**.

If the first noun phrase is indefinite (preceded by the indefinite article *a* or *an*), then repeat the indefinite article instead:

✗ Use the FILENAME statement to specify **a logical member name** and **member type**.

✓ Use the FILENAME statement to specify **a logical member name** and a **member type**.

⚠ Careful Analysis Is Required

Make sure you understand the effect that the articles *a*, *an*, and *the* have on modification. In this example from guideline 4.6.2, you saw that inserting *the* after *and* prevents *logical* from modifying *member type*:

✓ Use the FILENAME statement to specify **the logical member name** and the **member type**.

In other words, the *member name* is *logical*, but the *member type* is not. Let's look at another example:

▸ To clear the expression, use **the BACKSPACE** or **DELETE key**.

In the following revision, the author inserted *the* after *or*. By doing so, she prevented *BACKSPACE* from modifying *key*. In other words, her revision means that *BACKSPACE* is not a *key*:

⊘ To clear the expression, use **the BACKSPACE** or the **DELETE key**.

The author should have followed guideline 4.6.1 instead:

✓ To clear the expression, use **the BACKSPACE** key or the **DELETE key**.

Similarly, in the revised version of the next example, the *a* that follows *or* prevents *global* from modifying *definition table*:

✗ Create **a global** or **scorecard definition table** in your scorecard environment.

⊘ Create **a global** or a **scorecard definition table** in your scorecard environment.

Again, the author should have followed guideline 4.6.1 instead:

✓ Create **a global** definition table or a **scorecard definition table** in your scorecard environment.

4.6.3 Consider reversing the order of the noun phrases

In the examples in guideline 4.6.2, the modifier was modifying the first noun phrase but not a subsequent noun phrase. In addition, the modifier was preceded by an article (*a*, *an*, or *the*). If the modifier is not preceded by an article, then you must use a different revision strategy.

For example, the following sentence is ambiguous because *leading* can be interpreted either as modifying only *blanks* or as modifying both *blanks* and *semicolons*:

> ✗ Be careful when your input includes **leading blanks** and **semicolons**.

Suppose that *leading* modifies only *blanks*. You can resolve the ambiguity simply by reversing the order of the noun phrases:

> ✓ Be careful when your input includes **semicolons** and **leading blanks**.

4.6.4 Consider using an unordered list

If three or more coordinated noun phrases are at the end of a sentence, you can often resolve ambiguous modification by putting the noun phrases in a list:

> ✗ When you create a driver, you can specify **a basic**, **calculated**, **evenly assigned**, or **percentage driver**.

> ✓ When you create a driver, you can specify any of the following:
>
> - a basic driver
> - a calculated driver
> - an evenly assigned driver
> - a percentage driver

When you use this revision strategy, be sure to also follow guideline 3.8, "Use complete sentences to introduce lists."

4.6.5 Consider using a compound sentence

When a sentence contains two sets of conjoined noun phrases (often accompanied by the word *respectively*), the best solution might be to use a compound sentence. For example, consider the following sentence:

> ✗ The values of the **DBCSTYPE** and **DBCSLANG options** determine the **locale** and **character-set encoding**, respectively.

By repeating the subject and the verb, you can convert the sentence to a compound sentence that is easier to understand and translate:

> ✓ The value of **the DBCSTYPE option** determines **the locale**, and the value of the **DBCSLANG option** determines the **character-set encoding**.

4.6.6 Consider repeating a preposition

If the first noun phrase is part of a prepositional phrase, then consider repeating the preposition, in addition to whatever other revisions you make. For example, in the following sentence, the preposition *in* has two objects—*a SAS procedure* and *subroutine*:

✗ The stored query is executed when you use the view **in a SAS procedure** or **subroutine**.

As explained in guideline 4.6.2, by inserting an article (*a*) after *or*, you can make it clear that *SAS* does not modify *subroutine*:

✓ The stored query is executed when you use the view **in a SAS procedure** or a **subroutine**.

If you repeat the preposition as well, the added parallelism makes the sentence sound better:

✓+ The stored query is executed when you use the view **in a SAS procedure** or in a **subroutine**.

Because the sentence is now syntactically complete, it is also more suitable for machine translation.

Unfortunately, this revision strategy doesn't always work well stylistically. For example, in the following sentence, *fact* refers to a *fact table*. Therefore, you must repeat the head noun, as explained in guideline 4.6.1:

✗ Table Name is the name of each **fact** or **dimension table**.

✓ Table Name is the name of each **fact** table or **dimension table**.

If you also repeat the preposition *of* (along with the determiner *each*), the sentence sounds unnatural to many readers:

⊘ Table Name is the name of each **fact** table or of each **dimension table**.

Because the previous revision is clear to human readers and translators, there is no need to go to this extreme.

4.6.7 Consider inserting a translation note

Sometimes a revision that clarifies modification in conjoined noun phrases can seem awkward. For example, in the following sentence, suppose that the modifier *leading* modifies both *blanks* and *semicolons*:

✗ Be careful when your input includes **leading blanks** or **semicolons**.

If you repeat the modifier, as suggested in 4.6.1, the revised sentence sounds a bit odd:

> ? Be careful when your input includes **leading blanks** or leading **semicolons**.

If your intended meaning will be clear to your primary audience but not to your translators, then consider inserting a translation note instead of revising the sentence, as in this example:[4]

> ✓ Be careful when your input includes leading blanks **or** semicolons`<translationNote>"leading" modifies both "blanks" and "semicolons"</translationNote>`.

If you use machine-translation software, then syntactic clarity is more essential. In that case, you might choose to accept revisions that are less than perfect stylistically.

Standard or Idiomatic Phrases

Don't apply any of the above revision strategies to phrases that are standard or idiomatic. For example, *divide and conquer* is idiomatic even in non-technical discourse:

> ✓ In a **divide and conquer strategy**, a problem is divided into two or more subproblems that can be solved by the same technique.

However, use hyphens to show readers who are not familiar with the phrase that the entire phrase modifies *strategy*:

> ✓+ In a **divide-and-conquer strategy**, a problem is divided into two or more subproblems that can be solved by the same technique.

In the software domain, *query and reporting* is a standard technical term that should not be changed:

> ✓ A user of a **query and reporting application** can easily build a business report by using the parts of an information map as the building blocks for the report.

But again, hyphenate the phrase so that readers immediately recognize that *query-and-reporting* modifies *application*:

> ✓+ A user of a **query-and-reporting application** can easily build a business report by using the parts of an information map as the building blocks for the report.

---

[4] For more information about translation notes, see "Insert Explanations for Translators" in Chapter 1.

Chapter 5

Making Pronouns Clear and Easy to Translate

Introduction

Many organizations restrict the use of pronouns such as *I*, *we*, and *you* in certain types of documents. Similarly, so-called sexist pronouns such as *he* and *she* are typically avoided unless they refer to specific individuals. None of these pronouns poses a particular problem for global audiences. Therefore, this chapter focuses on pronouns that do pose problems for translators, non-native speakers, native speakers, or machine-translation software.

5.1 Make sure readers can identify what each pronoun refers to

Priority: HT2, NN2, MT2

English has several types of pronouns. However, of the *personal* pronouns in Table 5.1, only the highlighted third-person pronouns are potentially ambiguous and therefore problematic in technical documentation.

Table 5.1 Personal Pronouns

Person	Subject	Object	Possessive
1st person singular	I	me	my/mine
2nd person singular	you	you	your/yours
3rd person singular	he/she/it	him/her/it	his/her/hers/its
1st person plural	we	us	our/ours
2nd person plural	you	you	your/yours
3rd person plural	they	them	their/theirs

The highlighted pronouns are problematic because they sometimes have more than one possible referent. Unclear pronoun referents cause the following problems:

- In many languages, nouns have gender. In those languages, translations of *it*, *they*, and *them* differ depending on the number and gender of the pronoun's referent. For example, in French, Spanish, and other languages, the following sentence would be translated differently depending on whether the pronoun *it* refers to *the error* or to *your program*.
  - ▶ You must correct the error in your program before submitting **it** again.

In French, *it* would be translated as *le* if it refers to *program*, and as *la* if it refers to *error*:

- **T** Vous devez corriger l'erreur dans <u>votre programme</u> avant de **le** soumettre encore.
- **T** Vous devez corriger <u>l'erreur</u> dans votre programme avant de **la** soumettre encore.

If translators cannot determine what *it* refers to, then they must contact the author or another subject-matter expert for clarification. Obviously, pronouns that have more than one possible referent are also a problem for machine-translation software.

- Third-person pronouns that have more than one possible referent are often confusing even to native speakers. In the English sentence above, the noun *error* is in focus in most readers' minds when they encounter the pronoun *it*. (Many factors—including factors that have nothing to do with the text—affect what is in focus in a reader's mind as he or she is reading a particular sentence or paragraph.) But it's the program, not the error, that must be resubmitted. The lack of a clear, unambiguous referent for *it* is unsettling and confusing to many readers.

5.1.1 *It*

When *it* has more than one possible referent, you can often replace it with a form of the noun phrase that it refers to:

- **X** Once you define the <u>basic structure</u> of <u>your table</u>, enhancing **it** is easy.
- **✓** Once you define the basic structure of your table, enhancing **the table** is easy.
- **X** You cannot change <u>the REUSE= attribute</u> of <u>a compressed file</u> after **it** is created.
- **✓** You cannot change the REUSE= attribute of a compressed file after **the file** is created.

However, in some cases, this revision strategy violates the cardinal rule of Global English. For example, in the following sentence, *it* could be referring to either of two singular noun phrases, *syntax analyzer* and *RUN statement*:

- **?** When the <u>syntax analyzer</u> reaches the <u>RUN statement</u>, **it** passes the entire program to the compiler.

If you replace *it* with the appropriate noun phrase, the resulting sentence sounds unnatural:

? When the syntax analyzer reaches the RUN statement, **the syntax analyzer** passes the entire program to the compiler.

The fact that the revision sounds unnatural indicates that, to readers, the context made the pronoun's referent clear. Don't change the original sentence unless you can find a natural-sounding revision that resolves the potential ambiguity.

If there is any possibility that translators won't be sure what *it* refers to, then insert a translation note:[1]

✓ When the syntax analyzer reaches the RUN statement, **it** passes the entire program to the compiler`<translationNote>"it" refers to "syntax analyzer"</translationNote>`.

More about *it*

There are two other contexts in which *it* can be problematic for machine-translation software. Even if you are not using machine translation, you need to understand these contexts in order to identify what *it* refers to.

- Contexts in which *it* doesn't refer to anything:
  - If you store your definitions in external files, **it** is easy for you to share these definitions with others.
  - The Output Delivery System makes **it** easy to provide users with new formatting options.

  You don't need to change these non-referential uses of *it*, because they are not problematic for readers or translators. The first example, in which *it* is the subject of the verb *to be*, is seldom a problem even for machine-translation.

(*continued*)

[1] For more information about translation notes, see "Insert Explanations for Translators" in Chapter 1.

The second example, in which *it* is the object of a verb (*makes*), is more likely to be a problem for machine translation into some languages. However, that usage is very common, and it is often impossible to find suitable alternatives.

- Contexts in which *it* refers to an entire clause:
  - ▶ I couldn't believe **it**: The committee accepted my dissertation without requesting any modifications!
  - ▶ If **it** is not too much trouble, would you please bring me the newspaper?

  This usage is problematic for machine translation, but it rarely occurs in technical documentation.

5.1.2 *They*

Ambiguous uses of *they* are less common than ambiguous uses of *it*, but here is one example that many readers would find confusing:

✗ Critics of the plan argue that schools should teach children to speak Spanish better before **they** try to learn English.

You can easily revise the sentence as follows:

✓ Critics of the plan argue that schools should teach children to speak Spanish better before **teaching the children to speak English**.

This next example is more challenging:

? The rows in the data set must either be sorted by all the variables that you specify, or **they** must be indexed appropriately.

What is *they* referring to? Which of the two potential referents—*rows* or *variables*—can be indexed? For readers who are familiar with the subject matter, *rows* is probably in focus, so they interpret *they* as referring to *rows*. (That interpretation happens to be correct.) But would the referent be clear to a translator?

As you saw in one of the examples in guideline 5.1.1, replacing a potentially ambiguous pronoun with the noun that it refers to sometimes sounds awkward:

⊘ The rows in the data set must either be sorted by all the variables that you specify, or **the rows** must be indexed appropriately.

If you are confident that your main audience will not be confused by the ambiguity, then insert a translation note to tell translators what the pronoun is referring to:

✓ The rows in the data set must either be sorted by all the variables that you specify, or they `<translationNote>"they" refers to "rows"</translationNote>` must be indexed appropriately.

5.1.3 *Them*

In the following sentence, neither readers nor translators can be sure what *them* is referring to:

X Know the names of user interface elements, and use **them** correctly and consistently.

If the sentence is from a set of guidelines (for technical communicators, perhaps) that focus on using correct terminology, then *them* probably refers to *names*. If it is from guidelines for user-interface designers, then *them* probably refers to *user interface elements*.

? Know the names of user interface elements, and use **those names** correctly and consistently.

? Know the names of user interface elements, and use **those elements** correctly and consistently.

In a survey of a dozen technical writers and editors, even those who knew the context found the sentence confusing. The example and the survey illustrate two important points:

- It's easy to overlook ambiguities in your own writing—even if you are well versed in the Global English guidelines.
- Sometimes you should eliminate a potential ambiguity even if you think that the context makes your meaning clear.

In the next example, *them* could be referring to either *protected methods* or *labeled sections*:

X Labeled sections that are inside a method block must be coded as protected methods, and the statements that call **them** must be changed to method calls.

Again, all readers will be confused by this ambiguity, and translators will have to puzzle over the sentence for a long time before deciding (or perhaps guessing) which noun to regard as the pronoun's referent. As in the previous example, you can easily resolve the ambiguity by using a noun phrase instead of the pronoun:

✓ Labeled sections that are inside a method block must be coded as protected methods, and the statements that call **those sections** must be changed to method calls.

5.1.4 *Its*

Earlier you saw that, in many languages, pronouns agree in number and gender with the nouns that they refer to. Remember this sentence?

▸ You must correct the error in your program before submitting **it** again.

T Vous devez corriger l'erreur dans votre programme avant de **la** soumettre encore.

T Vous devez corriger l'erreur dans votre programme avant de **le** soumettre encore.

The first French translation is correct if the pronoun *it* refers to *error*, and the second translation is correct if *it* refers to *program.*

By contrast, the possessive pronoun *its* seldom poses a translation problem. (See the "Translating *its*" sidebar, below, for an explanation.) But when its referent is unclear, *its* impedes readability. In the following sentence, *its* has three possible referents: *numeric variable*, *character string*, and *formatted value*. To many readers, the sentence is confusing.

✗ If a numeric variable has a character string as a formatted value, **its** unformatted numeric value is transposed.

The following revision is easier to comprehend:

✓ If a numeric variable has a character string as a formatted value, **then the variable's** unformatted numeric value is transposed.

Even in the following example, where *its* can refer only to the singular noun phrase *bit mask*, the sentence is easier to comprehend if you repeat the noun phrase instead of using *its*:

? You can insert commas and blanks in a bit mask for readability without affecting **its** meaning.

✓ You can insert commas and blanks in a bit mask for readability without affecting **the bit mask's** meaning.

In the next example, *its* doesn't pose a problem for most readers. However, you can easily eliminate *its* with no loss of clarity:

? Piping enables the next process in the pipeline to read **its** input from the port instead of from a disk.

✓ Piping enables the next process in the pipeline to read input from the port instead of from a disk.

Translating *its*

Its is a possessive pronoun, not a subject or object pronoun. (See Table 5.1.) Therefore, in many other languages, *its* must agree in number and gender with the head noun that follows, not with the noun that it refers to. For example, in the following sentence, *its* refers to the noun phrase *Each program*, but it modifies the noun phrase *own interface*.

✓ Each program has **its** own interface.

In the following German translation, the word for *interface* is a feminine noun, *Schnittstelle*. Therefore, the German translations of *its* and *own* have feminine (*-e*) endings.

T Jedes Programm hat seine eigene Schnittstelle.

5.1.5 *Their*

With *their* as well as *its*, it is important to ensure that the referent of *their* is clear. Ambiguous uses of *their* occur less frequently than ambiguous uses of *its*, but here is one example:

✗ You can insert commas and blanks in the bit masks for readability without affecting **their** meanings.

In this example, the pronoun *their* has three possible referents: *commas*, *blanks*, and *bit masks*. The following revision is easier to comprehend:

✓ You can insert commas and blanks in the bit masks for readability without affecting **the meanings of the bit masks**.

5.2 Don't use *this*, *that*, *these*, and *those* as pronouns

Priority: HT2, NN3, MT2

Use these words only as adjectives (followed by nouns) so that their referents are clear.

It's easy to determine when these words are being used as pronouns. Since pronouns take the place of nouns, simply look for places where one of these words is *not* followed by a noun, as in this example:

▶ Apply all the Before Rules manual adjustments. **These** are listed on the `Before Rules` tab of the Adjustments window.

By contrast, when they are followed by a noun, these same words function as adjectives:

▶ Apply all the Before Rules manual adjustments. **These adjustments** are listed on the `Before Rules` tab of the Adjustments window.

Like the pronouns that were discussed in 5.1.1–5.1.3, these pronouns can pose a translation problem, because in many languages they must agree in number and gender with the nouns that they are referring to.

In the following example, the use of *these* in the second sentence is not problematic for human translators, because the first sentence contains a plural noun (*adjustments*) that *these* is clearly referring to. However, the sentence could be problematic for machine-translation software.

✗ Apply all the manual adjustments. **These** are listed on the `Rules` tab of the Adjustments window.

✓ Apply all the manual adjustments. **These adjustments** are listed on the `Rules` tab of the Adjustments window.

As is often the case, there is a more concise alternative:

✓+ Apply all the manual adjustments that are listed on the `Rules` tab of the Adjustments window.

More often, the preceding sentence doesn't contain a noun that translators and machine-translation software can use in their translations.

✗ Some data might be aggregated while you are using the OLAP page. **This** can result in poor performance.

In this case, it would not be too difficult for a French translator to determine that the noun *situation*, which happens to be the same in French as in English, would be suitable:

T (French) … **Cette situation** peut entraîner une dégradation des performances.

Nevertheless, it is better to include the noun in the English sentence:

✓ Some data might be aggregated while you are using the OLAP page. **This situation** can result in poor performance.

The following example illustrates how difficult and time-consuming it can be for translators to translate *this*, *that*, *these*, and *those* when they are used as pronouns. The sentence that contains *this* must be revised substantially in order to make it translatable:

X In subsequent passes through the data, the rows are read from the spool file rather than being reread from the original data sources. **This** guarantees that each pass through the data processes the exact same information.

✓ In subsequent passes through the data, the rows are read from the spool file rather than being reread from the original data sources. Each pass through the data is therefore guaranteed to process the exact same information.

More about *this*, *that*, *these*, and *those*

When these words are used without a following noun, they usually occur at the beginning of a sentence. However, they can also occur later in a sentence, as the second *this* in the following example illustrates:

X By default, users of your application can use the OVERRIDE command to enter a value that exceeds the specified maximum. **This** allows the value to be stored in the data set. You can prevent **this** by specifying in the `Override on errors` field that overriding is not allowed.

✓ By default, users of your application can use the OVERRIDE command to enter a value that exceeds the specified maximum. **The OVERRIDE command** allows the value to be stored in the data set. **To prevent users from overriding the specified maximum value**, specify in the `Override on errors` field that overriding is not allowed.

(continued)

The second of the three sentences probably isn't even necessary:

✓+ By default, users of your application can use the OVERRIDE command to enter a value that exceeds the specified maximum. **To prevent users from overriding the specified maximum value**, specify in the `Override on errors` **field** that overriding is not allowed.

As you apply the Global English guidelines, remember to look for opportunities to eliminate non-essential information and to reduce word counts.

5.3 Don't use *which* to refer to an entire clause

Priority: HT2, NN3, MT2

When you use *which* as a relative pronoun, make sure it refers to a specific noun phrase, as in the following example:

✓ Suppose you want to print the contents of the Retail.Europe table, **which** contains data about European customers.

Don't use *which* to refer to an entire clause, as in the following example:

X Blade servers take up less room than racks, **which** is good because as your server needs increase, your data center size doesn't.

Such sentences are often difficult to translate, because many languages don't have relative pronouns that can be used in this manner.

When you revise such sentences, be careful not to violate guideline 5.2, "Don't use *this*, *that*, *these*, and *those* as pronouns." The following revision is a violation of guideline 5.2 because *that* is being used as a pronoun:

⦸ Blade servers take up less room than racks. **That** is good, because as your server needs increase, you won't need to increase the size of your data center.

In order to eliminate this undesirable use of *which*, you must revise the sentence substantially:

✓ Blade servers take up less room than racks. As a result, you won't need to increase the size of your data center as your server needs increase.

Here are some other examples that illustrate a variety of revision strategies:

✗ The Netgear ReadyNAS NV+ offers a lot of features and great performance, **which** is good, because it's expensive.

✓ The Netgear ReadyNAS NV+ is expensive, but it offers a lot of features, as well as great performance.

✗ Python doesn't support circular references, **which** is a problem because circular references often occur in code where the programmer has no control over the retain count.

✓ Python's lack of support for circular references is a problem, because circular references often occur in code where the programmer has no control over the retain count.

✗ The IMPORT command skips over symbolic links, **which** is a problem because many of the required files are installed as symbolic links.

✓ The IMPORT command skips over symbolic links. **This behavior** causes a problem, because many of the required files are installed as symbolic links.

✗ The distances between quasars and the telescopes that detect them are large, **which** means the light from the quasars shifts toward the red end of the spectrum.

✓ Because the distances between quasars and the telescopes that detect them are large, the light from the quasars shifts toward the red end of the spectrum.

✗ Socket inheritance limits the number of ports that are used for connections to firewalls, **which** increases the security of private networks.

✓ Socket inheritance increases the security of private networks by limiting the number of ports that are used for connections to firewalls.

This next sentence uses *which* correctly, because *which* refers to *strong aversion*, not to the entire preceding clause:

✓ In Germany there is a strong aversion to the use of genetically altered foodstuffs, **which** has only intensified as a result of recent studies.

However, because there is so much distance between *which* and its referent, it is better to divide the sentence and repeat the noun:

✓+ In Germany there is a strong aversion to the use of genetically altered foodstuffs. **That aversion** has only intensified as a result of recent studies.

More about *which*

Guideline 5.3 pertains to contexts in which *which* is being used as a relative pronoun. Don't try to apply the guideline to contexts in which *which* fills some other grammatical role.

For example, sometimes *which* is an adjective:

- No one really knows **which** dinosaur was the smartest.

And sometimes it is an interrogative pronoun:

- Of the three games, **which** was the hardest to create and produce?

Chapter 6

Using Syntactic Cues

Introduction

Generally speaking, a syntactic cue is any element or aspect of language that helps readers identify parts of speech and analyze sentence structure. For example, prefixes, suffixes, articles (*a*, *an*, and *the*), prepositions, punctuation marks, and even word order are all syntactic cues.

Syntactic cues are so powerful that they enable readers to make grammatical sense out of the following lines from Lewis Carroll's "Jabberwocky," even though the content words are nonsense:

> 'Twas brillig, and the slithy toves
> Did gyre and gimble in the wabe.

Even readers who have never heard of syntactic cues know that *toves* is a noun, because it ends in *-s* and is preceded by the article *the*. They know that *slithy* is an adjective, because it ends in *-y* (a typical adjectival suffix) and because it occurs between an article and a noun. And they know that *gyre* and *gimble* are verbs because of the presence of the auxiliary verb *did*.

In this book, the term *syntactic cue* refers only to syntactic cues that can be omitted without making a sentence, clause, or phrase ungrammatical or incomprehensible. If you have taken courses in technical writing, then many of the words that your instructors taught you to omit for the sake of brevity are syntactic cues. For example, in the following sentence, the word *that* is a syntactic cue:

✓ The user configuration file **that** you specify is executed along with the system configuration file **that** your installation uses.

But removing the two occurrences of *that* from the sentence makes the sentence more difficult to read and understand—especially for non-native speakers of English:

✗ The user configuration file you specify is executed along with the system configuration file your installation uses.

Sentences from which syntactic cues have been omitted are often ambiguous and difficult to translate, in addition to being difficult to comprehend. So forget what you were taught, and put syntactic cues back into your writing! There are plenty of other ways to reduce word counts without depriving your global audience of words that are helpful or even essential. (See Appendix A, "Examples of Content Reduction.")

A detailed discussion of the research that supports the use of syntactic cues is provided in Appendix D, "Improving Translatability and Readability with Syntactic Cues." Therefore, this chapter provides only brief explanations of why each type of syntactic cue is necessary or useful.

 **Reminder: The Cardinal Rule of Global English**

The cardinal rule of Global English is especially applicable to syntactic cues. Use syntactic cues with discretion. Be careful not to change meaning or emphasis, and don't use so many syntactic cues that your text sounds unnatural to native speakers of English. Most important, always consider whether some other type of revision, or some additional revision, would be better than merely inserting a syntactic cue.

The Cardinal Rule of Global English

Don't make any change that will sound unnatural to native speakers of English.

Corollary

There is almost always a natural-sounding alternative if you are creative enough (and have enough time) to find it!

6.1 Don't use a telegraphic writing style

Priority: HT1, NN1, MT1

A telegraphic writing style, in which multiple articles or other syntactic cues are omitted, is not suitable for global audiences. As the following examples illustrate, this style of writing is often difficult even for native speakers to read and comprehend:

- ✗ LABEL option not supported for file format.
- ✓ **The** LABEL option **is** not supported for **this** file format.
- ✗ SHEET name ignored if conflict occurs with RANGE name specified.
- ✓ **The** SHEET name **is** ignored if it conflicts with **the** RANGE name **that was** specified.

Even in sentences that are not obviously telegraphic, you should look for opportunities to insert articles and other determiners:

> ✗ The following features improve usability of the product for users who have disabilities.

> ✓ The following features improve **the** usability of the product for users who have disabilities.

However, don't violate guideline 2.8, "Use *the* only with definite nouns."

6.2 In a series of noun phrases, consider including an article in each noun phrase

Priority: HT2, NN3, MT2

If some of the noun phrases in a series are definite and others are indefinite, then include the appropriate article in each noun phrase:

> ✗ **The** base dictionary, user-defined dictionary, and document-specific dictionary can all be used by the spell-checking software.

> ✓ **The** base dictionary, **a** user-defined dictionary, and **a** document-specific dictionary can all be used by the spell-checking software.

Similarly, if some indefinite noun phrases start with vowels (requiring *an*), and others start with consonants (requiring *a*), then include the appropriate article in each noun phrase:

> ✗ If you choose to create **an** applet, application, console application, or servlet, you can set the following options.

> ✓ If you choose to create **an** applet, **an** application, **a** console application, or **a** servlet, you can set the following options.

For an explanation of definite and indefinite nouns, see Kohl (1990).

6.3 Use *that* with verbs that take noun clauses as complements

Priority: HT2, NN2, MT2

Whenever you use any form of the verbs or verb phrases *assume*, *be sure*, *ensure*, *indicate*, *make sure*, *mean*, *require*, *specify*, *suppose*, and *verify*, consider inserting *that* to make the sentence structure more explicit. In this context, *that* helps readers (especially non-native speakers) and machine-translation software recognize that a noun clause follows. In the following example, the noun clause is enclosed in brackets:

✗ A check mark next to the table name indicates [the table has been selected].

✓ A check mark next to the table name indicates **that** [the table has been selected].

Here are some other sentences in which you can improve readability and translatability by inserting the syntactic cue *that*:

✗ It is important to ensure the LDRTBLS value is large enough to contain your X client.

✓ It is important to ensure **that** the LDRTBLS value is large enough to contain your X client.

✗ Suppose the data is all numeric data and it is currently stored in a raw data file.

✓ Suppose **that** the data is all numeric data and **that** it is currently stored in a raw data file.

Be Sure and *Make Sure*

Both *be sure* and *make sure* should often be followed by *that*, but they are not exact synonyms.

Be sure that is not appropriate when you are issuing a command. When you are telling readers to do something, use *be sure to*:

✗ **Be sure that** you brush your teeth regularly.

✓ **Be sure to** brush your teeth regularly.

(*continued*)

Be sure that is appropriate only in other, non-imperative contexts:

✓ Without checking the date of my virus definitions file, I could not **be sure that** I was protected against the NoamChomsky virus.

When you are telling readers to ensure or verify something, use *make sure that*, *ensure that*, or *verify that*:

X **Be sure that** all elements can be used without a mouse.

✓ **Make sure that** all elements can be used without a mouse.

✓ **Ensure that** all elements can be used without a mouse.

X **Be sure that** the USCPI table contains summarized data.

✓ **Verify that** the USCPI table contains summarized data.

Don't use *make sure to* in any context. That usage is not standard English:

⊘ **Make sure to** upload your images to the server.

✓ **Be sure to** upload your images to the server.

⊘ **Make sure to** vote.

✓ **Be sure to** vote.

6.4 Use *that* in relative clauses

Priority: HT2, NN2, MT2

In English, there is a type of restrictive relative clause[1] in which the relative pronoun is the object of the clause's verb and can be omitted. In Global English, you should almost always include these relative pronouns in order to make the sentence structure explicit.

X The file you selected is displayed in the frame on the right.

✓ The file **that** you selected is displayed in the frame on the right.

X The page you requested could not be located.

✓ The page **that** you requested could not be located.

[1] See guideline 4.3, "Clarify what each relative clause is modifying," for explanations of restrictive and non-restrictive relative clauses.

As usual, don't insert this syntactic cue if doing so would make the sentence sound unnatural:

► The only other thing you can do is keep yourself well hydrated.

? The only other thing **that** you can do is keep yourself well hydrated.

► The only thing I know for sure is that my little brother is an alien.

? The only thing **that** I know for sure is that my little brother is an alien.

And as always, consider alternative revisions. In the following example, the revised sentence might briefly confuse some readers because the prepositional phrase *to their original values* could be modifying either *change* or *to return*:

✗ Remember to return any global settings you change to their original values.

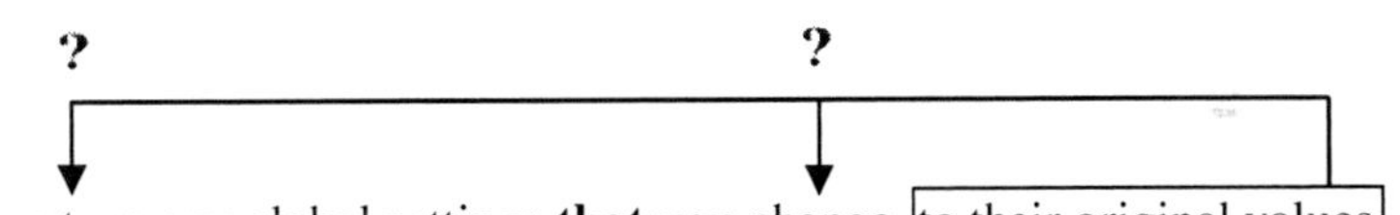

? Remember to return any global settings **that** you change to their original values.

If you move the prepositional phrase closer to *return*, the sentence sounds strange to many native speakers:

⊘ Remember to return [to their original values] any global settings **that** you change.

In this case, the best solution is a complete revision that doesn't even include a relative clause:

✓+ If you change any global settings, be sure to change them back to their original values.

 Ungrammatical Use of *that*

Be careful not to insert *that* in contexts where it is ungrammatical. In the revision of the following sentence, *that* does not have a legitimate grammatical function:

- X Statements execute in the order they are written.
- ⊘ Statements execute in the order **that** they are written.

In order to produce a grammatical sentence, *in which* must be inserted instead:

- ✓ Statements execute in the order **in which** they are written.

If this type of revision makes the sentence sound unnatural, then revise the sentence in some other way:

- X The fundamental way computers work has not changed.
- ⊘ The fundamental way **that** computers work has not changed.
- ? The fundamental way **in which** computers work has not changed.
- ✓ The fundamentals of computers have not changed in decades.

- X The only way the differential would matter would be if there were ties in several categories.
- ⊘ The only way **that** the differential would matter would be if there were ties in several categories.
- ? The only way **in which** the differential would matter would be if there were ties in several categories.
- ✓ The differential would matter only if there were ties in several categories.

6.5 Clarify which parts of a sentence are being joined by *and* or *or*

Priority: HT2, NN2, MT2

As you saw in guideline 4.5, when the conjunctions *and* and *or* join noun phrases, the scope of modification is often ambiguous. But *and* and *or* can also lead to another type of ambiguity, called ambiguous *scope of conjunction*.

The term *scope of conjunction* refers to the parts of a sentence that are being joined by *and* or *or*. Ambiguous *scope of conjunction* can sometimes be humorous or silly:

▶ Do not dip your bread or roll in your soup.

If you interpret *or* as joining two verb phrases, then the scope of conjunction is broad:

? Do not [dip your bread] **or** [roll in your soup].

If you interpret *or* as joining two noun phrases, then the scope of conjunction is narrow:

? Do not dip [your bread] **or** [roll] in your soup.

You can eliminate the ambiguity by inserting *your* before *roll*:

✓ Do not dip your bread **or** <u>your</u> roll in your soup.

This form of the possessive pronoun *your* is always followed by a noun.[2] Thus, *your* is a syntactic cue that makes it clear that *roll* is a noun rather than a verb.

In technical documents, ambiguous scope of conjunction can be confusing or even dangerous. For example, the following sentence is confusing to all readers:

X The application can use the window to establish a dialog with the user **and** format text responses.

Here are two possible interpretations of the sentence:

? The application [can use the window to establish a dialog with the user] **and** [<u>**can**</u> format text responses].

? The application can use the window [to establish a dialog with the user] **and** [<u>**to**</u> format text responses].

---

[2] Adjectives, adverbs, and other nouns can intervene between a possessive pronoun and a head noun, as in <u>*your*</u> *extremely expensive, high-definition plasma* <u>*TV*</u>.

The second revision of this sentence is correct. But without the syntactic cue, neither a reader, a human translator, nor machine-translation software can be sure about the correct interpretation.

Not Just a Matter of Style

Many authors and editors recognize that using syntactic cues following *and* or *or* adds a pleasing parallelism to their sentences:

> ✗ The coordinates correspond to the top of the vertical slide bar **and** the left edge of the horizontal slide bar.

> ✓ The coordinates correspond [**to** the top of the vertical slide bar] **and** [**to** the left edge of the horizontal slide bar].

But as you saw in the previous examples, parallelism is not just a stylistic issue. The syntactic cues that create parallel grammatical structures are often essential for clear communication.

Examples of Syntactic Cues That Resolve Scope-of-Conjunction Problems

Depending on the context, many different syntactic cues can be used to resolve scope-of-conjunction problems. Here are several examples:

Prepositions

> ✗ DQCleanse enables you to standardize your data by building schemes from your data **and** applying those schemes to your data.

> ✓ DQCleanse enables you to standardize your data **by building** schemes from your data **and by applying** those schemes to your data.

The infinitive marker *to*

> ✗ You cannot use the DATASETS procedure to change the values of rows, change or delete columns, **or** change the type or length of variables.

> ✓ You cannot use the DATASETS procedure **to change** the values of rows, **to change or delete** variables, **or to change** the type or length of variables.

That

- ✗ Typically, the message indicates that a variable has been identified incorrectly **or** the raw data file contains some invalid data values.
- ✓ Typically, the message indicates **that** a variable has been identified incorrectly **or that** the raw data file contains some invalid data values.

Auxiliary verbs

- ✗ Because the expression produces a numeric value, TOTAL is defined as a numeric variable **and** assigned a default length of 8.
- ✓ Because the expression produces a numeric value, TOTAL **is defined** as a numeric variable **and is assigned** a default length of 8.

Modal verbs

- ✗ Character values must be enclosed in quotation marks **and** be in the same case as in the data set.
- ✓ Character values **must be enclosed** in quotation marks **and must be** in the same case as in the data set.

Subordinating conjunctions

- ✗ Position the statement in the program so that it checks the condition as soon as it is logically possible **and** unnecessary statements do not execute.
- ✓ Position the statement in the program **so that** it checks the condition as soon as it is logically possible**, and so that** unnecessary statements do not execute.
- ✗ If a single lookup might take several seconds **and** many clients might request service simultaneously, then the server must be able to handle multiple clients concurrently.
- ✓ **If** a single lookup might take several seconds**, and if** many clients might request service simultaneously, then the server must be able to handle multiple clients concurrently.

Tips for Clarifying Scope of Conjunction

Tip 1: Consider making additional revisions

Keep in mind that merely inserting a syntactic cue is not always the best way to resolve a scope-of-conjunction ambiguity. In the following example, the second revision, in which the sentence has been rearranged and revised further, conveys the intended meaning better than the first revision:

✗ The ID variables list is a list of variables whose values rarely change **and** are useful for identification.

✓ The ID variables list is a list of variables whose values rarely change **and** **<u>that</u>** are useful for identification.

✓+ The ID variables list is a list of variables **<u>that</u>** are useful for identification because their values rarely change.

The next revision is more concise and eliminates the need for the syntactic cue. However, moving *the variables* to the beginning of the sentence changes the emphasis. The author must decide whether the change in emphasis is acceptable:

? The variables in the ID variables list are useful for identification because their values rarely change.

Tip 2: Be careful not to accidentally distort your meaning

Make sure you understand which parts of a sentence are being joined by *and* or *or*. Otherwise, you might distort the meaning of the sentence by inserting the wrong syntactic cue. In the next example, inserting *to* after *and* is wrong because pointer controls are not used for naming variables:

✗ The values for SALARY begin in column 19, so use the pointer control to point to column 19 **and** name the variable.

⊘ The values for SALARY begin in column 19, so use the pointer control to point to column 19 **and** **<u>to</u>** name the variable.

A subject-matter expert would revise the sentence correctly as follows:

✓ The values for SALARY begin in column 19, so use the pointer control to point to column 19**<u>,</u> and** **<u>then</u>** name the variable.

Note: If an editor or peer reviewer inserts syntactic cues, the author must review the edits carefully to ensure that all ambiguities are resolved correctly.

(*continued*)

Tip 3: Syntactic cues are not always necessary or desirable

In general, don't insert a syntactic cue if the parts of a sentence that are being joined by *and* or *or* are so close together that there is no potential for misreading or ambiguity:

✓ There is no need to re-sort a table when rows are modified **or** added.

⊘ There is no need to re-sort a table when rows are modified **or** **are** added.

✓ The text substitution that this activity produces is complete before the program text is compiled **and** executed.

⊘ The text substitution that this activity produces is complete before the program text is compiled **and** **is** executed.

However, if two different grammatical constructions are being joined, then insert a syntactic cue even if the two parts are close together. In the following example, *known* is an adjective, and *built* is a participle that forms the passive verb *are built*:

X The tags are known **and** built into the browser.

✓ The tags [are known] **and** [**are** built into the browser].

6.6 Revise past participles

Priority: HT2, NN2, MT2

Overview of Past Participles

A past participle is a form of a verb that usually ends in *-ed* or *-d*, as in *distinguish**ed*** and *describ**ed***. However, the past participles of some verbs are formed by making internal spelling changes or by adding other endings, as in *begun*, *done*, *paid*, and *written*.[3] The past participle is the form of a verb that is used to form the present perfect tense.

In the following table, the past participle is in bold in the third column. As you can see, the past participle is often the same as the past-tense form of a verb:

[3] For a list of irregular verbs, search the Web for the exact phrase *list of English irregular verbs*.

Table 6.1 Present, Past, and Present Perfect of Some Common English Verbs

Present	Simple Past	Present Perfect
I begin	I began	I have **begun**
I describe	I described	I have **described**
I do	I did	I have **done**
I follow	I followed	I have **followed**
I simplify	I simplified	I have **simplified**
I understand	I understood	I have **understood**

6.6.1 Revise past participles that follow and modify nouns

Past participles that follow nouns are problematic for non-native speakers as well as for machine-translation software because, as you saw in Table 6.1, a past participle is often identical to the past-tense form of a verb. When you expand the participle into a relative clause, the sentence structure is more explicit and is easier to analyze.

In some contexts, you can convert the participle to a relative clause by inserting *that* and making other small adjustments to the sentence. For example, in the following sentence, the past participial construction, *used by the NETFLOW procedure*, can be converted to a relative clause, *that the NETFLOW procedure uses*:

✗ This is the same algorithm **used** by the NETFLOW procedure.

✓ This is the same algorithm **that** the NETFLOW procedure **uses**.

In other contexts, you must insert *that* plus some form of the verb *to be* (*that is, that are, that was, that were, that has been, that have been*, and so on), as in this example:

✗ The Checklist feature provides a method of cataloging tasks **performed** during routine procedures.

✓ The Checklist feature provides a method of cataloging tasks **that are performed** during routine procedures.

However, keep in mind that the present-tense form of *to be* is not always appropriate:

✗ The template **used** for the episode is overlaid on each waveform.

⊘ The template **that is used** for the episode is overlaid on each waveform.

✓ The template **that was used** for the episode is overlaid on each waveform.

✗ The Cardiac Compass report includes an entry for every defibrillation therapy **delivered**.

⊘ The Cardiac Compass report includes an entry for every defibrillation therapy **that is delivered**.

✓ The Cardiac Compass report includes an entry for every defibrillation therapy **that has been delivered**.

This guideline is also applicable to sentences in which *not* separates the noun from the participle:

✗ Any argument **not enclosed** in angle brackets is a required argument.

✓ Any argument **that is not enclosed** in angle brackets is a required argument.

⚠ Careful Analysis Is Required

The past participles in the following sentences cannot be expanded into relative clauses without making the sentences ungrammatical:

- The Sequence flag always has the two low bits **set** to 0.
- Leave the `Authenticate` option **checked**.

Don't expand a past participle if doing so would make the sentence sound unnatural to native speakers or would change the emphasis in the sentence.

6.6.2 Revise past participial phrases that follow commas

If a past participial phrase follows a comma, make the sentence structure more explicit by changing the participial phrase to a relative clause:

✗ The Life Peerage, **created by the Life Peerages Act of 1958**, is often regarded as inferior to the Hereditary Peerage.

✓ The Life Peerage, **which was created by the Life Peerages Act of 1958**, is often regarded as inferior to the Hereditary Peerage.

An even better revision strategy is to put the main clause and the participial phrase in separate sentences. Single-clause sentences are often easier to translate than sentences that contain embedded relative clauses:

✓+ The Life Peerage was **created by the Life Peerages Act of 1958**. It is often regarded as inferior to the Hereditary Peerage.

Here is another example that illustrates the two revision strategies:

- ✘ The list of the top 50 cities for fall allergens, **compiled by the Allergy Foundation of America,** shows the Raleigh-Durham metropolitan area as number 2.
- ✓ The list of the top 50 cities for fall allergens, **which was compiled by the Allergy Foundation of America,** shows the Raleigh-Durham metropolitan area as number 2.
- ✓+ The list of the top 50 cities for fall allergens shows the Raleigh-Durham metropolitan area as number 2. **The list was compiled by the Allergy Foundation of America.**

As always, consider whether other, more concise, revisions are possible:

- ✓+ In a list **that was compiled by the Allergy Foundation of America**, the Raleigh-Durham metropolitan area is number 2.

6.7 Revise adjectives that follow nouns

Priority: HT2, NN2, MT2

When an adjective follows a noun, consider expanding the adjective into a relative clause:

- ✘ Intervals **shorter** than the VF interval are counted by the VF event counter.
- ✓ Intervals **that are shorter** than the VF interval are counted by the VF event counter.
- ✘ The RTF option produces output **suitable** for Microsoft Word reports.
- ✓ The RTF option produces output **that is suitable** for Microsoft Word reports.

Relative clauses are more syntactically explicit than adjectives that follow nouns. Therefore, relative clauses are easier for non-native speakers of English to comprehend and for machine-translation systems to analyze.

Remember that instead of merely inserting syntactic cues, you should always be looking for opportunities to make sentences more concise:

- ✗ The data **available** in the episode log includes the following types of data:
- ✓ The data **<u>that is</u> available** in the episode log includes the following types of data:
- ✓+ The episode log includes the following types of data:

Related Guideline

- 4.2.3, "If a prepositional phrase modifies a noun phrase, consider expanding it into a relative clause"

6.8 Use *to* with indirect objects

Priority: HT3, NN2, MT2

Whenever you use the verb *give* or *assign* with both a direct object and an indirect object, consider using the word *to* to make the indirect object more syntactically explicit. In the following examples, the direct objects are underlined, and the indirect objects are italicized:

- ✗ A label gives *a variable* <u>a more informative name</u>.
- ✓ A label gives <u>a more informative name</u> ***to*** *a variable.*
- ✗ Before printing a date, you usually assign *it* <u>a format</u>.
- ✓ Before printing a date, you usually assign <u>a format</u> ***to*** *it.*

Notice that when you use the word *to* in this context, the indirect object typically follows the direct object and receives greater emphasis. If the added emphasis seems inappropriate or sounds unnatural to you, then don't follow this guideline.

Other Verbs to Consider

This guideline is probably also applicable to other verbs that can take both direct objects and indirect objects. Here is a partial list of such verbs that might be used in technical documentation: *award*, *bring*, *buy*,[4] *deny*, *grant*, *guarantee*, *hand*, *lease*, *leave*, *lend*, *loan*, *offer*, *owe*, *pass*, *pay*, *promise*, *read*, *recite*, *rent*, *sell*, *send*, *serve*, *show*, *take*, *teach*, *tell*, *throw*, *toss*, *wire*, and *write*.

You might want to collect examples of sentences that contain these verbs in order to determine how frequently this guideline could be applied to each verb.

6.9 Consider using *both . . . and* and *either . . . or*

Priority: HT3, NN3, MT3

When appropriate, consider using the correlative conjunctions *both . . . and* and *either . . . or* rather than just *and* and *or*. Pairs of correlative conjunctions improve readability by making it easier for readers to see which parts of a sentence are being conjoined. They also facilitate human translation and machine translation.

However, in order for correlative conjunctions to add clarity rather than confusion, you must place them in the correct locations. When you use *both . . . and* or *either . . . or*, make sure that they frame similar grammatical constructions. The following sentence is incorrect because *not found in the autocall library* doesn't have the same grammatical structure as *it could not be opened*:

✗ The macro was **either** [not found in the autocall library] **or** [it could not be opened].

In the corrected version of the sentence, both *either* and *or* are followed by subjects and verb phrases:

✓ **Either** [the macro was not found in the autocall library] **or** [the macro could not be opened].

[4] With the verb *buy*, use the preposition *for* to mark the indirect object.

Examples of *both . . . and*

✗ The lexicographer must specify the syntactic category of the word **and** the categories of the phrases that it can take as complements.

✓ The lexicographer must specify **both** the syntactic category of the word **and** the categories of the phrases that it can take as complements.

✗ Don't include a structure tag **and** variables or definitions in the same declaration.

✓ Don't include **both** a structure tag **and** variables or definitions in the same declaration.

Examples of *either . . . or*

✗ This model has a push button **or** a slide switch with which you select the resolution.

✓ This model has **either** a push button **or** a slide switch with which you select the resolution.

✗ If the range is not dense, then convert your integers to strings **or** do a binary search on an array of possible values.

✓ If the range is not dense, then **either** convert your integers to strings **or** do a binary search on an array of possible values.

⚠ Careful Analysis Is Required

Note that *both* and *either* are not always conjunctions. In the following sentences, they are adjectives. When they are used as adjectives, they don't frame similar grammatical constructions even if they happen to be followed by *and* or *or*.

- No waiver by **either** party of any breach or default shall be deemed to be a waiver of any subsequent breach or default.
- Individuals might claim tax credits for **both** types of solar-energy systems and for using heat pumps that have SEER ratings of 12 or higher.

6.10 Consider using *if . . . then*

Priority: HT3, NN3, MT3

If *if* introduces a conditional clause, then consider beginning the following clause with *then*. The *then* in an *if . . . then* construction improves readability partly by providing semantic reinforcement. That is, it reinforces the idea of a condition being followed by a result or consequence. It also serves as a syntactic cue, clearly signaling the beginning of the main clause.

- ✗ **If** you have not assigned a logical name to the data file, specify the physical filename in the statement that refers to the file.
- ✓ **If** you have not assigned a logical name to the data file, **then** specify the physical filename in the statement that refers to the file.
- ✗ **If** you used the SDL script to build the Xmingw compiler, the script should have built the full set of tools, including the C++ compiler.
- ✓ **If** you used the SDL script to build the Xmingw compiler, **then** the script should have built the full set of tools, including the C++ compiler.

If inserting *then* doesn't make the sentence sound better to you, then don't do it. For example, when the verb in the *if* clause is in present tense, the use of *then* often seems to be less appropriate:[5]

- ? **If** the plastic ratchet straps are dirty, **then** it's very difficult to loosen them because the dirt causes the microlock to seize.
- ? **If** variations arise in the female line, **then** they are not likely to be transmitted exclusively in that line.
- ? **If** you want to filter the data that an information map can access, **then** use a stored process to perform that filtering.

[5] No one has ever precisely identified the linguistic contexts in which using *then* after an *if* clause is appropriate or helpful without sounding unnatural. The tenses of the *if* clause and the main clause are not the only factors.

6.11 Make each sentence syntactically and semantically complete

Priority: HT1, NN2, MT1

In addition to omitting the specific types of syntactic cues that have already been addressed in this chapter, authors occasionally omit other words or phrases that they consider optional in certain contexts. But syntactic or semantic incompleteness makes sentences more difficult for non-native speakers to comprehend and for some types of machine-translation systems to analyze. It can also make the sentences more difficult for human translators and for other readers who are not familiar with the subject matter.

The widespread misuse of *that*, which was mentioned in guideline 6.4, is one example of some authors' tendency to be syntactically incomplete:

- ✗ The contingency table lists eye and hair color in the order **that** they appear in the data set.
- ✓ The contingency table lists eye and hair color in the order **in which** they appear in the data set.

Sometimes authors omit so much semantic content that a sentence is quite ambiguous. For example, it is impossible for anyone other than the author to know which interpretation of the following sentence is correct:

- ✗ **Accessing the return code** is operating-system specific.
- ? The way in which return codes are accessed is operating-system specific.
- ? Return codes are accessed on some operating systems, but not on others.
- ? Return codes are accessed differently on some operating systems than on others.

Similarly, most readers and translators would not fully understand what *in hexadecimal* means in the following sentence. The author omitted too much of the semantic content:

- ✗ Examine the sample file, checking for the correct syntax **in hexadecimal**.
- ✓ Examine the sample file, checking for the correct syntax **in the hexadecimal representation of the heading**.

In the following examples, the intended meanings are less difficult to discern. Nevertheless, the ideas in these sentences should be stated more explicitly.

- ✗ Space is tracked and reused in the compressed file according to the REUSE value when the file was created, not when you add and delete records.
- ✓ Space is tracked and reused in the compressed file according to the REUSE value **that was in effect** when the file was created, not **according to the REUSE value that is in effect** when you add and delete records.

- ✗ If you are **uncertain which** table you need, browse through all the tables.
- ✓ If you are uncertain **about which** table you need, browse through all the tables.
- ✓+ If you are **not sure which** table you need, browse through all the tables.

- ✗ Because the Java applet is executing locally, it can provide a much richer interactive environment than a CGI program **can**.
- ✓ Because the Java applet is executing locally, it can provide a much richer interactive environment than a CGI program **can provide**.

- ✗ If employees have planned to do any work in an Oasis application this weekend that can be delayed until Monday, they should consider **doing so**.
- ✓ Any employees who planned to use an Oasis application this weekend should consider **waiting until Monday**.
- ✓+ If you were planning to use an Oasis application this weekend, please consider **waiting until Monday**.

Chapter 7

Clarifying -ING Words

Introduction

In this book, the term *-ING word* refers to a word that is formed by adding *-ing* to the root form of a verb. By this definition, words that are not formed from verbs (for example, *during*, *string*, *wing*, and *nothing*) are not -ING words. In addition, words that end in *-ings*, such as *settings* and *writings*, are not -ING words.

-ING words are problematic for the following reasons:

- -ING words can fill many different grammatical roles, and many languages don't have equivalent constructions. Therefore, -ING words sometimes confuse non-native speakers of English.
- Even native speakers use -ING words ungrammatically or don't punctuate -ING words correctly in certain contexts. Faulty grammar and punctuation pose problems for human translators and for machine-translation software.
- In some contexts, -ING words are ambiguous and confusing even to native speakers.

Nevertheless, -ING words are useful and often essential parts of the English language. Instead of banning them in all contexts, this chapter focuses on contexts in which -ING words are frequently ambiguous or ungrammatical and in which reasonable alternatives exist.

Any discussion of -ING words has to include grammatical terms and concepts that many authors are not familiar with. If you need a more complete explanation of the terms, or if you want a deeper understanding of the grammar, then see "The Grammar of -ING Words" on page 151. The more arcane terms are also defined in the glossary.

7.1 Revise -ING words that follow and modify nouns

Priority: HT2, NN2, MT2

If an -ING word immediately follows *and modifies* a noun, then either expand it into a relative clause or find some other way of eliminating it.

✗ Move the certificate authority to a new server **running** on a domain controller.

✓ Move the certificate authority to a new server **that is running** on a domain controller.

You can use this same revision strategy for passive -ING constructions, in which the -ING word *being* is followed by a past participle. In the following example, *being defined* modifies the noun *capabilities* and can be expanded into *that are being defined*:

✗ Include comments to explain the capabilities **being defined**.

✓ Include comments to explain the capabilities **that are being defined**.

If active voice is suitable for a particular context, then use that instead:[1]

✓+ Include comments to explain the capabilities **that you are defining**.

Because the -ING words in the revised sentences above are now preceded by *that* plus a form of the verb *to be*, their grammatical roles are easier for non-native speakers, machine-translation systems, and even native speakers to recognize.

As with past participles (guideline 6.6.1), two types of relative clauses—one that includes some form of the verb *to be*, and one that doesn't—can be used for revising -ING words that follow and modify nouns. Choose the type of clause that is suitable for the context.

For example, in the sentence below, a revision that uses the relative clause *that are consisting* is not appropriate:

✗ The relationships among data items are expressed by tables **consisting** of columns and rows.

⊘ The relationships among data items are expressed by tables **that are consisting** of columns and rows.

The appropriate revision uses *that consist*:

✓ The relationships among data items are expressed by tables **that consist** of columns and rows.

As usual, other types of revisions are sometimes required. In the following example, *specifying* modifies *VBAR statement*:

✗ The following example shows a VBAR statement **specifying** that the frequency be displayed inside the bars.

If you expand *specifying* into either type of relative clause, the revised sentences don't sound good:

? The following example shows a VBAR statement **that is specifying** that the frequency be displayed inside the bars.

---

[1] See guideline 3.6, "Limit your use of passive voice."

? The following example shows a VBAR statement **that specifies** that the frequency be displayed inside the bars.

Here are some shorter alternatives that sound better:

✓ In the following example, a VBAR statement specifies that the frequency be displayed inside the bars.

✓+ In the following example, a VBAR statement causes the frequency to be displayed inside the bars.

Here is another example in which an alternative revision strategy is preferable:

✗ If you try to update the file, you receive an error message **indicating** that you do not have permission to write to the file.

? If you try to update the file, you receive an error message **that indicates that** you do not have permission to write to the file.

✓ If you try to update the file, an error message **informs you that** you do not have permission to write to the file.

When Not to Apply This Guideline

Remember that this guideline is applicable only to -ING words that follow *and modify* nouns. Let's look at some examples of -ING words that follow but do not modify nouns.

In the following sentence, neither *sharing* nor *trading* modifies *file*:

▶ Peer-to-peer file **sharing** is not the same as file **trading**.

In that sentence, *file sharing* and *file trading* are noun phrases that each consist of the noun *file* plus a gerund. (See "Gerund, Adjective, or Noun?" on page 151 for a discussion of gerunds.) If you apply guideline 7.1, the sentence becomes nonsensical:

⊘ Peer-to-peer file **that is sharing** is not the same as file **that is trading**.

Applying guideline 7.1 to the next sentence also results in nonsense, because *looking* does not modify *hour*:

▶ The user spent an hour **looking** for a solution.

⊘ The user spent an hour **that was looking** for a solution.

(*continued*)

In some contexts, applying guideline 7.1 to an -ING word that does not modify the preceding noun distorts the meaning of the sentence. In the following sentence, if you expand *using* into a relative clause, the implication is that the folder is using the `rmdir` command:

▶ Delete the folder **using** the `rmdir` command.

⊘ Delete the folder **that is using** the `rmdir` command.

To eliminate any possible confusion, you could revise the sentence as follows:

✓ **Use** the `rmdir` command **to delete** the folder.

Another alternative is to insert a comma, as discussed in guideline 7.4:

✓ Delete the folder**, using** the `rmdir` command.

Here is another example in which applying guideline 7.1 distorts the meaning of the sentence:

▶ I saw the clerks **destroying** evidence.

⊘ I saw the clerks **who were destroying** evidence.

What the author meant was this:

✓ I saw the clerks **as they were destroying** evidence.

7.2 Revise -ING words that follow certain verbs

Priority: HT3, NN2, MT2

In general, if an -ING word follows any of the verbs *begin*, *start*, *continue*, or *prefer*, then eliminate the -ING word by changing it to an infinitive.[2]

X DATAPRO/2000 **continues processing** program statements after it repairs the data set.

✓ DATAPRO/2000 **continues to process** program statements after it repairs the data set.

[2] You can also apply this guideline to the verbs *hate*, *love*, and *like*. However, those verbs are seldom used in technical documentation.

✗ El Niño conditions **started developing** in the central Pacific Ocean during August.

✓ El Niño conditions **started to develop** in the central Pacific Ocean during August.

However, if an -ING word follows the *infinitive* form of one of these verbs, then don't revise it. In the revisions of the following sentences, the two consecutive infinitives sound awkward to many readers:

✓ The mayor asked all residents **to start conserving** water.

✗ The mayor asked all residents **to start to conserve** water.

✓ After computing the exponent and the simple moving average, you are ready **to begin calculating** the exponential average.

✗ After computing the exponent and the simple moving average, you are ready **to begin to calculate** the exponential average.

7.3 Revise dangling -ING phrases

Priority: HT1, NN3, MT2

Like other ungrammatical constructions, dangling -ING phrases pose significant problems for human translators and for machine-translation software. Revise dangling -ING phrases by supplying a subject for the -ING word. In this revision strategy, you usually have to replace the -ING word with a different form of the same verb as well:

✗ **When <u>using</u> a Wi-Fi hot spot**, your laptop's file sharing does not need to be enabled.

✓ **When <u>you use</u> a Wi-Fi hot spot**, your laptop's file sharing does not need to be enabled.

Don't revise -ING phrases that are grammatical:

✓ **When <u>registering</u> a control-window class**, you must specify how many window words will be associated with that class.

Now let's examine these examples in more detail. In the first example, the -ING phrase, *When using a Wi-Fi hot spot*, is subordinate to the main clause of the sentence, which begins with *your laptop's file sharing*:

✗ **When using a Wi-Fi hot spot**, your laptop's file sharing does not need to be enabled.

In order to be grammatical, the subject of *using* must be the same as the subject of the main clause. Because *your laptop's file sharing* is not *using a Wi-Fi hot spot*, the *When using* phrase is a dangling -ING phrase.

In order to translate such sentences, translators often must first determine what the real subject of the -ING phrase should be.[3] The author should save translators that trouble by replacing *using* with an explicit subject (*you*) and a different form of the verb (*use*):

✓ **When you use a Wi-Fi hot spot**, your laptop's file sharing does not need to be enabled.

Because many authors are unaware of this particular point of English grammar, -ING phrases that begin with the following subordinating conjunctions are often ungrammatical:

- *after*
- *before*
- *by*
- *prior to*
- *since*
- *until*
- *when*
- *while*

[3] In some languages and contexts, a dangling -ING phrase can be translated using a grammatical structure that does not require an explicit subject. However, the goal in this chapter is to eliminate all -ING words that are frequently ambiguous or ungrammatical and for which reasonable alternatives exist.

The other example that you saw above is clear and correct, because the subject of the main clause, *you*, is also the subject of *registering*:

✓ **When registering a control-window class**, you must specify how many window words will be associated with that class.

Grammatical -ING phrases like this one are a natural part of the English language. From a Global English standpoint, there is no reason to eliminate -ING words from such contexts.

In the next example, the -ING phrase (*while updating*) is part of the clause that begins with *If an error occurs*. The -ING phrase dangles because the subject of *updating* is not the same as the subject of the *If* clause (*an error*).

✗ If an error occurs **while updating a VSAM file**, the operating system might be able to recover the file and repair some of the damage.

In other words, the -ING phrase is ungrammatical because the following interpretation is incorrect:

⊘ If an error occurs **while an error is updating a VSAM file**, the operating system might be able to recover the file and repair some of the damage.

Notice that this time the subject of the *main* clause (*the operating system*) doesn't affect whether the -ING phrase is grammatical or not. Whenever an -ING phrase is part of a subordinate clause (in this case, the *If* clause), the subject of the -ING word must be the same as the subject of the subordinate clause, not the subject of the main clause.

Nevertheless, as in the previous example, a translator might have to assign a subject to *updating* in order to translate it. Since there are few other noun phrases in the sentence that could be the subject of *updating*, the translator might choose to use *the operating system* as the subject. Again, that interpretation is incorrect:

⊘ If an error occurs **while the operating system is updating a VSAM file**, the operating system might be able to recover the file and repair some of the damage.

A translator would likely have to consult the author in order to be sure how to interpret the dangling -ING phrase. In this case, the following interpretation is correct:

✓ If an error occurs **while you are updating a VSAM file**, the operating system might be able to recover the file and repair some of the damage.

Implied Subjects of Imperative Verbs

In order to identify dangling -ING phrases, you must correctly identify the subject of each clause. Keep in mind that the subject of an imperative (command) verb is always *you*, even though the subject is not expressed:

- [You] **Compare** the pre-VF and pre-VT rhythms to the patient's normal sinus rhythms.
- To accept the licensing agreement, [you] **click** `Accept`.

In the following sentence, the verb in the main clause is *measure*, and its implied subject is *you*. Because *you* (the reader) is also the subject of *measuring*, the sentence is grammatical and should not be revised:

✓ **When measuring the pacing values**, measure between the tip and the ring of each bipolar sensor.

In the next example, the -ING phrase *setting a sorting priority* is the subject of the main clause. Since there is no mention of who or what is doing the setting, there is no agent (subject) for the -ING words *creating* or *editing*, either:

X Previously, setting a sorting priority was available only **when creating or editing a report**.

The solution is to put human agents into the sentence:

✓ Previously, **you** could set a sorting priority only when **you** were creating or editing a report.

If the agent of an action is unknown or unimportant, then it is better to use a passive verb than to allow an -ING phrase to dangle. However, use active voice whenever possible.

In the following example, the -ING phrase dangles because the subject of the main clause is not the subject of *applying*:

X Some of the heavily skewed variables are more normally distributed **after applying a log transformation**.

In the following revision, *a log transformation* becomes the subject of the passive verb *is applied*:

✓ Some of the heavily skewed variables are more normally distributed **after a log transformation is applied**.

However, this active-voice revision is better if the reader is the agent of the action:

✓+ Some of the heavily skewed variables are more normally distributed **after <u>you apply</u> a log transformation**.

7.4 Punctuate -ING phrases correctly

Priority: HT2, NN2, MT2

When an -ING phrase immediately follows and modifies a main clause (as opposed to modifying just the preceding noun, as discussed in guideline 7.1), precede it with a comma, as in these examples:

✓ The box moves as you move the pointer, **enabling** you to position the viewer.

✓ The Nomad II lacks internal memory, **relying** instead on flash-memory cards that slide into the player's battery chamber.

If the -ING phrase follows a noun, then the meaning is different depending on whether the comma is present. For example, the following sentences have different meanings:

▸ Restart the servers **following** these guidelines:

▸ Restart the servers, **following** these guidelines:

In the first sentence, the absence of a comma can cause readers to briefly misinterpret *following* as modifying only the noun *servers*. If that were the intended meaning, then guideline 7.1 would make the meaning explicit:

▸ Restart the servers **following** these guidelines:

? Restart the servers **that follow** these guidelines:

In the second version of the sentence, the presence of a comma makes it clear that *following* modifies the entire main clause (*Restart the servers*):

▸ Restart the servers, **following** these guidelines:

In other words, the comma prevents readers and machine-translation software from possibly misinterpreting or being confused by the -ING phrase.

Here are some other examples in which a necessary comma has been omitted:

- ✗ Roll-up arithmetic takes different forms **depending** on the relevant account type and the relevant Frequency member.
- ✓ Roll-up arithmetic takes different forms**, depending** on the relevant account type and the relevant Frequency member.
- ✗ UTF-8 supports all of the world's languages **including** those that use non-Latin1 characters.
- ✓ UTF-8 supports all of the world's languages**, including** those that use non-Latin1 characters.
- ✗ Thirty million tape cartridges could store nearly one million times the contents of the Library of Congress **assuming** an average capacity of 400GB per cartridge.
- ✓ Thirty million tape cartridges could store nearly one million times the contents of the Library of Congress**, assuming** an average capacity of 400GB per cartridge.
- ✗ The GNPV total equals the NPV of total economic production costs **excluding** land costs.
- ✓ The GNPV total equals the NPV of total economic production costs, **excluding** land costs.

If the -ING phrase follows a verb, then the meaning is not affected, but the lack of a comma is non-standard and is jarring to some readers. For example, in software documentation, the comma should not be omitted after the verbs *open* and *appear*:

- ✗ The package **opens displaying** the process flow diagram.
- ✓ The package **opens, displaying** the process flow diagram.
- ✗ A dialog box **appears asking** you to verify that you want to condense the selected nodes into a subdiagram.
- ✓ A dialog box **appears, asking** you to verify that you want to condense the selected nodes into a subdiagram.

⚠ *Following* Can Be a Preposition

If you can replace *following* with *after*, then *following* is being used as a preposition. In that case, inserting a comma would be incorrect:

✓ When you create or modify an expression, the Operators menu automatically **appears following** the selection of an operand.

⊘ When you create or modify an expression, the Operators menu automatically **appears, following** the selection of an operand.

However, in this case, you can easily eliminate *following* and make a few other improvements at the same time:

✓+ When you create or modify an expression, the Operators menu **appears after you select** an operand.

7.5 Hyphenate -ING words in compound modifiers

Priority: HT2, NN2, MT2

Always hyphenate compound modifiers that include -ING words:

✗ Click `File Name` or `File Reference` to specify your preferred file **referencing** method.

✓ Click `File Name` or `File Reference` to specify your preferred file-**referencing** method.

✗ The platform suite includes a scheduler and a load **sharing** facility.

✓ The platform suite includes a scheduler and a load-**sharing** facility.

✗ A template **sizing** cursor appears in the Graph window.

✓ A template-**sizing** cursor appears in the Graph window.

You should usually hyphenate other compound modifiers as well. See guideline 8.7.1, "Consider hyphenating noun phrases," for more information.

7.6 Eliminate unnecessary -ING phrases and -ING clauses

Priority: HT2, NN2, MT1

If an -ING phrase makes a sentence excessively long and complex, then find a way to divide and simplify the sentence:

- ✗ With this choice, a complete roll-up through the Source hierarchy occurs before the allocation process**, incorporating** all adjustments that were laid down earlier in the adjustment process. (1 sentence, 28 words)
- ✓ With this choice, a complete roll-up through the Source hierarchy occurs before the allocation process. All adjustments that were laid down earlier in the adjustment process **are incorporated**. (2 sentences, 29 words)

Even when a sentence that contains an -ING phrase is short, you can sometimes find a suitable revision that doesn't include an -ING phrase:

- ✗ Enter 14 as the third split point, **making** sure you select `Add Branch`.
- ✓ Enter 14 as the third split point. **Be** sure to select `Add Branch`.

Consider eliminating -ING *clauses* as well. In an -ING clause, the -ING word is the subject of a verb:

- ✗ **Uploading** the entries from memory might be useful if you do not need to use the resident functions.
- ✓ If you do not need to use the resident functions, **you can upload** the entries from memory.
- ✗ **Selecting** any node causes the text of that node to appear in a Text Entry control.
- ✓ **When you select** a node, the text of that node appears in a Text Entry control.
- ✗ **Using** AppleTalk access lists simplifies network management.
- ✓ AppleTalk access lists simplify network management.

Beware of Revisions That Violate Other Guidelines

Beware of revisions that violate other Global English guidelines. Consider the following example:

✓ The second argument has been set to NULL**, indicating** that the future amount is to be calculated.

You might be tempted to revise the sentence as follows:

X The second argument has been set to NULL**. This indicates** that the future amount is to be calculated.

However, this revision violates guideline 5.2, "Don't use *this*, *that*, *these*, and *those* as pronouns." Instead, you could either revise the sentence as follows or leave it unchanged:

✓ The second argument has been set to NULL**. The NULL value indicates** that the future amount is to be calculated.

7.7 Revise ambiguous -ING + noun constructions

Priority: HT3, NN3, MT3

Except in titles and headings, -ING + noun constructions are rarely problematic for humans. In most contexts, syntactic cues make -ING + noun constructions unambiguous.[4] Where these constructions *are* ambiguous, other information in the surrounding context usually makes the meaning clear.

[4] See "Some Contexts in Which -ING Words Are Unambiguous" on page 155.

Nevertheless, occasionally it is not clear whether the -ING word in an -ING + noun construction is a gerund, an adjective, or a noun.[5] Consider the following example:

> **?** CCPC is a practical tool that has been designed for **certifying** agencies.

If *certifying* is a gerund, then it refers to an action. In that case, the sentence could be rephrased as follows:

> ✓ CCPC is a practical tool that has been designed for **the purpose of certifying** agencies.

If *certifying* is an adjective, then it is an attribute or characteristic of *agencies*. In that case, the sentence could be rephrased as follows:

> ✓ CCPC is a practical tool that has been designed for **use by certifying** agencies.

Translators need to know which interpretation is correct. If the original sentence is not revised, and if the phrase *certifying agencies* doesn't occur in some other context in which the -ING word is unambiguous, then translators will have to either query the author or make an educated guess.

As the following sentence illustrates, this type of ambiguous construction can also confuse readers:

> **?** A job can include **scheduling** metadata that enables the program to run at a specified date and time.

Many readers initially interpret *scheduling* as a gerund (action). To avoid the ambiguity, the author could revise the sentence as follows:

> ✓ A job can include the **scheduling** of metadata to enable the program to run at a specified date and time.

But the author probably intended *scheduling* to be an adjective. In that case, the author could use the following revision:

> ✓ A job can include metadata **that schedules** the program to run at a specified date and time.

---

[5] For an explanation of how to distinguish between gerunds, adjectives, and nouns, see "Gerund, Adjective, or Noun?" on page 151.

Ambiguous -ING + Noun Constructions in Headings and Titles

Ambiguous -ING + noun constructions occur much more frequently in headings and titles (including titles of windows and dialog boxes) than in running text. Here are some examples that have multiple interpretations:

? **Defining Characteristics**

- gerund: the action of defining characteristics
- adjective: characteristics that define something

? **Changing Import Options**

- gerund: the action of changing import options
- adjective: import options that are changing

? **Tracking Software Publishers**

- gerund: the action of keeping track of companies that publish software
- noun: publishers of software that is used for tracking something

? **Editing Options**

- gerund: the action of editing options
- noun: options that pertain to editing something

? **Encoding Names**

- gerund: the action of encoding names
- noun: names of (character-set) encodings

In documentation, the context that follows a heading or title usually makes it clear which meaning is intended. However, if most of the -ING words in your headings or titles are gerunds, then try to avoid using -ING words that are adjectives or nouns.

(*continued*)

The examples above in which the -ING word is most likely a noun could be revised as follows:

✓ Publishers of Manuscript-Tracking Software[6]

✓ Options for Editing

✓ Names of Encodings

Note: In software localization, translators often have to translate user-interface text—including titles of windows and dialog boxes—without having access to the software. Because they start localizing the software while the documentation is still under construction, they don't have access to the documentation, either. Therefore, ambiguities in software are often more of a problem than ambiguities in documentation.

7.8 Revise ambiguous *to be* + -ING constructions

Priority: HT3, NN3, MT3

Like -ING + noun constructions, -ING words that follow a form of *to be* are rarely ambiguous or confusing. However, consider the following example:

X Another way of conserving storage space **is compressing** files.

Many readers are initially confused by this sentence, in which the -ING word *compressing* is a gerund. To make it clear that *compressing* is a gerund, you can revise the problem sentence as follows:

✓ You can also conserve storage space **by compressing** files.

---

[6] The example was the title of an article about software that helps freelance writers keep track of which publishers they have submitted articles to.

Readers are confused by the original sentence because it is much more common for *to be* + -ING to form a progressive verb form, as in these examples:

- ▶ The form has been submitted to the next level and **is awaiting** review.
- ▶ If you **are defining** columns for a new table, you must specify the column attributes.
- ▶ If you **will be installing** software on any UNIX system, you should also create an Installer account.

The next example poses the same kind of problem:

- X One of the first tasks you should consider **is creating** summary statistics for each of the variables.

In the following revision, the syntax is simpler and is not subject to misreading:

- ✓ Before performing other tasks, consider **creating** summary statistics for each of the variables.

In the next example, *are reporting* could be misinterpreted as a progressive verb form:

- ▶ A balancing rule looks for improper inequalities within the accounts of all organizations that **are reporting** entities.

However, if *reporting entity* is a technical term that readers have encountered elsewhere, then readers probably will not be confused by it. In that case, just be sure to include the term in a glossary for translators.

Predicate Adjectives

An -ING word that follows a form of *to be* can also be a *predicate adjective*— an adjective that is on the other side of the verb from the noun that it is modifying.

- ▶ The dot-com boom of the 1990s **was exciting** while it lasted. [*exciting* modifies *The dot-com boom*]
- ▶ The music was bad, the acoustics sucked, and the atmosphere **was depressing**. [*depressing* modifies *the atmosphere*]

However, most of the -ING words that can be used as predicate adjectives pertain to emotions. Therefore, they rarely occur in technical documentation. In any case, human readers are not likely to be confused by this construction.

The Grammar of -ING Words

Understanding the grammar of -ING words can be a challenge. Grammar books and Web sites don't agree on how to classify or explain -ING words, and most explanations are oversimplified.

However, a thorough understanding of -ING grammar is essential for anyone who wants to master Global English. In addition to helping you recognize when -ING words *are* ambiguous, this understanding will help you recognize when they are *not* ambiguous. Because the contexts in which -ING words are not ambiguous far outnumber the contexts in which they are, you will save yourself a lot of trouble by learning about this aspect of English grammar.

Gerund, Adjective, or Noun?

Table 7.1 provides several examples of -ING words that can be used either as gerunds or as adjectives or nouns.[7] As you can see, gerunds typically convey actions or activities, whereas adjectives and nouns describe characteristics or attributes.

Table 7.1 -ING Words Used as Gerunds and as Adjectives or Nouns

Gerund (action)	Adjective or Noun (characteristic)
Developing photographic film requires patience and skill.	**Developing** nations need access to credible, independent scientific and technological information.
Growing hemp has been proposed as an alternative to **growing** tobacco.	**Growing** deficits threaten the U.S. economy.
Decreasing defects in our software has improved our customer-retention rate.	**Decreasing** oil-industry profits led to a decline in the Dow Jones Industrial Average.

(*continued*)

[7] Some -ING words are rarely used as adjectives in technical documentation. (Examples include *attempting*, *deriving*, *imagining*, *missing*, and *thanking*.) Some -ING words are rarely used as gerunds. (Examples include *exciting* and *intervening*.) However, many -ING words can be gerunds, adjectives, or nouns.

Table 7.1 (*continued*)

Gerund (action)	Adjective or Noun (characteristic)
Charging consumers for services that used to be free could backfire on the banking industry.	A **charging** rhinoceros rammed the bus ahead of us.
Montgomery got into the business of **cleaning** industrial-waste sites when he was 16.	The new store carries a huge selection of **cleaning** products.
Answering questions about English grammar is challenging, but fun.	**Answering** machines are rapidly becoming obsolete.

Adjectives and nouns are in the same column of the table because when they precede nouns they can both be considered modifiers. They identify a subset of the following noun. For example, *growing deficits* refers only to the subset of deficits that are growing, and *answering machines* refers only to the subset of machines that perform the function of answering.

Elsewhere in this chapter, the difference between -ING words that are used as adjectives and -ING words that are used as nouns is not important. However, for the sake of completeness, here is an explanation:

Suppose you use the term *moving boxes* to refer to boxes in which you pack items when you are changing your place of residence. In this case, *moving* is not a gerund because you are not talking about the action or process of moving boxes. It also is not an attribute or characteristic of the boxes. If your neighbor says "Do you have any moving boxes that I can have?", he is not asking for boxes that are moving. In this context, *moving* is simply a noun, and *moving boxes* is a compound noun. Similarly, in the phrases *greeting cards* and *welding equipment*, *greeting* and *welding* are nouns, not adjectives or gerunds.

Here is another way to distinguish between -ING words that are adjectives and -ING words that are nouns:

- The phrase *developing nations* can be rephrased as *nations that are developing*. -ING words that can be rephrased in this manner are adjectives.
- In most contexts, the phrase *cleaning products* doesn't refer to the action or process of cleaning products; hence, *cleaning* is not a gerund. In addition, *cleaning products* cannot be rephrased as *products that are cleaning*, so *cleaning* is not an adjective. By the process of elimination, *cleaning* is a noun. In this case, the phrase could be expressed as *products that are used for cleaning* or as

products that pertain to or somehow involve cleaning. That type of revision probably works for most -ING + noun phrases in which the -ING word is a noun.

In the phrase *answering machines, answering* is a bit more difficult to figure out. It is probably best classified as a noun, because it would be more reasonable to rephrase the construction as *machines that are used for answering* than as *machines that are answering. Singing telegram* is another example in which the -ING word is a noun.

More about Gerunds

Even though they convey actions or activities, gerunds can fill many of the same grammatical roles as nouns.

Subjects:

- **Gardening** was Ray's passion.

Direct objects:

- Susie had always hated **swimming**, but the shipwreck renewed her interest in the activity.

Indirect objects:

- The editor's corrections gave **nitpicking** a whole new meaning.

Objects of prepositions:

- By **swimming** parallel to the shore, Susie was eventually able to escape the rip current.

Gerunds also have some characteristics of verbs. For example, gerunds frequently have their own objects:

- **Specifying** the system password gives you full administrative access.

  (the system password = object)

Even a gerund that is the object of a verb or preposition can have its own object:

- There are several ways of **verifying** the validity of a passenger's identification.

  (the validity = object)

-ING Clauses

In an *-ING clause*, the -ING word is a gerund that is the subject of a verb:

- If the application defines a default toolbar for its application window, then [**clicking** the Restore Defaults button restores the settings for that toolbar].
- [**Applying** the WEST argument to the spatial data causes the longitudes to be negated] when the data is read in.

An -ING clause can begin with a subordinating conjunction such as *when* or *if*. In that case, the -ING clause is a subordinate clause rather than a main clause:

- [When **editing** is complete], the action class should display the portlet again.
- Consider indexing your data [if **generating** reports quickly is your most important consideration].

Remember that -ING words are sometimes modifiers (nouns or adjectives) rather than gerunds. In the following sentence, *computing resources*—not just *computing*—is the subject of *are represented*. Thus, *computing* is a modifier, not a gerund, and the clause that begins with *computing* is not an -ING clause:

- On a metadata server**, computing** resources are represented by metadata objects.

-ING Phrases

In an *-ING phrase*, the -ING word is also a gerund. However, it is not the subject of a verb.

As you saw in guideline 7.1, -ING phrases sometimes follow and modify nouns:

- A single arrow [**pointing** to the right] always takes users to the next 20 rows.

In that case, you should usually expand the -ING phrase into a relative clause:

- A single arrow **that points** to the right always takes users to the next 20 rows.

An -ING phrase can also be the object of a preposition:

- There are several ways of [**verifying** the validity of a passenger's identification].

An -ING phrase can modify an entire clause. In that case, a comma should separate the -ING phrase from the clause that it is modifying :

- [**Depending** on the content type], the portal either displays the item or launches a new application in your browser window.
- Complete the Save dialog box, [**selecting** Template as the type of report].

An -ING phrase can contain an infinitive or even a finite verb. However, in an -ING phrase, the -ING word is not the subject of the verb. In the following example, *to position* is an infinitive:

- The box moves as you move the pointer, [**enabling** you to position the viewer].

In this example that you saw earlier, the subject of the verb *takes* is *arrow*, not *pointing*:

- A single arrow [**pointing** to the right] always takes users to the next 20 rows.

In the next example, the -ING phrase is followed by a subordinate clause. The subordinate clause has its own subject (*you*) and verb (*make*). Thus, the -ING word (*viewing*) is not the subject of *make*:

- It is helpful to make changes incrementally, [**viewing** the results of each change] before you make the next change.

Here is an example in which the -ING phrase is followed by a relative clause (*that were left by other windows*).

- Redraw the window, **eliminating** any shadows that were left by other windows.

The relative pronoun *that* is the subject of *were left*. So again you see that in contrast to an -ING *clause*, in an -ING *phrase*, the -ING word is not the subject of a verb.

As you saw in guideline 7.3, -ING phrases can start with subordinating conjunctions such as *when*, *after*, and *before*. If the -ING phrase *precedes* the main clause, then it is followed by a comma.

- When **initializing** lists, you can enclose the list items in either braces or brackets.
- After **entering** the required information, click `View Report`.

If the subordinate -ING phrase *follows* the main clause, then it is not preceded by a comma:

- The search examines all of the values in each column before **proceeding** to the next row.

Some Contexts in Which -ING Words Are Unambiguous

Now that you know more about the grammar of -ING words, it will be easier for you to understand why the majority of -ING words are not ambiguous or problematic for human translators or other readers. As noted earlier, if you understand the unambiguous contexts, then you will be better able to focus your attention on recognizing the -ING words that *are* potentially ambiguous or confusing.

Earlier in this chapter you saw some contexts in which syntactic cues make -ING words unambiguous:

- In guideline 7.8 you saw that -ING words that follow a form of *to be* are rarely ambiguous or confusing.
- In guidelines 7.4 and 7.5 you saw that commas and hyphens can make -ING words unambiguous.

Thus, commas, hyphens, and the verb *to be* are all syntactic cues.

Table 7.2 shows some other contexts in which syntactic cues make -ING words unambiguous. In the Context column, the syntactic cues are in bold.

Determiners are one of these syntactic cues. Articles (*a*, *an*, *the*) are the most common type of determiners, but determiners also include numbers, quantifiers (*many*, *much*, *some*, *several*, *a few*, and so on), demonstratives (*this*, *that*, *these*, and *those*), and possessive pronouns (*my*, *your*, *their*, and so on).

Table 7.2 Contexts in Which -ING Words Are Never Ambiguous

Context	Example	Grammatical Role
-ING word + **determiner or adjective** + noun	**Retiring** an account is similar to **removing** an account.	gerund (action)
Determiner or adjective + -ING word + noun	The **training** service was started successfully.	modifier (adjective or noun)
-ING word + plural noun + **inflected verb**	**Adding** segments **is** easier because very little reconfiguration is required. Adding **machines** **are** not yet obsolete.	gerund or modifier, depending on the verb's inflection

The following sections discuss these unambiguous contexts in more detail.

-ING word + determiner or adjective + noun

When an -ING word precedes one or more determiners or adjectives and a noun, it is unambiguous. In this context, an -ING word is always a gerund. Remember that gerunds convey actions or processes, whereas adjectives convey attributes or characteristics.

In the following examples, the determiners and adjectives are underlined:

- ✓ **Resizing** the component causes components that contain text objects to overlap.
- ✓ **Keeping** other risk factors to a minimum can help prevent infection.
- ✓ **Pretreating** the Y-79 cells failed to alter homologous PAC(1) receptor desensitization.
- ✓ **Converting** local-variable accesses into stack accesses can improve the performance of stack-based programs.
- ✓ **Filtering** an Alphanumeric Category in a List Table

Determiner or adjective + -ING word + noun

If a determiner or adjective precedes an -ING word + noun construction, the -ING word is always a modifier.

In the following examples, the determiners and adjectives are underlined:

- ✓ Choose a **starting** point and take the appropriate action.
- ✓ If the current **filtering** choices are not acceptable, you can use the Table Data dialog box to assign data items to different functions.
- ✓ Local conventions can include specific **formatting** rules for dates, times, and numbers.
- ✓ The investigation showed that usability was a **determining** factor in the popularity of Indian Web sites.

Table 7.3 shows how preceding an -ING word with a determiner or adjective makes the -ING word unambiguous:

Table 7.3 -ING Phrases to Which Articles or Determiners Have Been Added

Ambiguous	**Interpretations**	**Unambiguous**	**Interpretation**
resizing components	**gerund**: the action or process of resizing components **modifier**: components that are used for resizing something	the resizing components	*resizing* can only be a modifier (in this case, a noun)
stabilizing processes	**gerund**: the action or process of stabilizing processes **modifier**: processes that are stabilizing something	intrinsic stabilizing processes	*stabilizing* can only be a modifier (in this case, an adjective)
refurbishing equipment	**gerund**: the action or process of refurbishing equipment **modifier**: equipment that is used for refurbishing something	expensive refurbishing equipment	*refurbishing* can only be a modifier (in this case, a noun)[8]
measuring cups	**gerund**: the action or process of measuring cups **modifier**: cups that are used for measuring	plastic measuring cups	*measuring* can only be a modifier (in this case, a noun)

[8] A gerund interpretation is possible but unlikely. In that interpretation, the phrase would mean *equipment that is used for expensive refurbishing* in contrast to equipment that is used for inexpensive refurbishing.

-ING word + plural noun + inflected verb

Even when an -ING word + plural noun is not preceded by a determiner or adjective, the context usually makes it clear whether the -ING word is a gerund or an adjective. For example, in the following sentence, the fact that the verb *causes* is third-person singular tells you that *Reporting* is the subject. When the verb agrees with the -ING word, the -ING word can only be a gerund:

▶ Reporting violations causes the agency a lot of headaches.

By contrast, if *Reporting* is merely modifying *violations*, then *violations* is the subject, and the verb is therefore third-person plural:

▶ Reporting violations cause the agency a lot of headaches.

As Table 7.4 shows, for the tenses or verb forms in which the verb has a different form for a singular subject than it has for a plural subject, an -ING word + plural noun combination is unambiguous. The verb acts as a syntactic cue that tells you whether the -ING word is a modifier or a gerund.

Table 7.4 Tenses or Verb Forms in Which an -ING + Plural Noun Are Unambiguous

Present tense:

✓ Reporting violations causes the agency a lot of headaches.

✓ Reporting violations cause the agency a lot of headaches.

Any tense in which *have* is an auxiliary verb:

✓ Reporting violations has caused the agency a lot of headaches.

✓ Reporting violations have caused the agency a lot of headaches.

✓ Reporting violations has been causing the agency a lot of headaches.

✓ Reporting violations have been causing the agency a lot of headaches.

Any tense in which *to be* is an auxiliary verb:

✓ Reporting violations is causing the agency a lot of headaches.

✓ Reporting violations are causing the agency a lot of headaches.

✓ Reporting violations was causing the agency a lot of headaches.

✓ Reporting violations were causing the agency a lot of headaches.

(continued)

Table 7.4 (*continued*)

Constructions in which *do* is an auxiliary verb:

✓ Reporting violations does cause the agency a lot of headaches.

✓ Reporting violations do cause the agency a lot of headaches.

✓ Does reporting violations cause the agency a lot of headaches?

✓ Do reporting violations cause the agency a lot of headaches?

Table 7.5 shows the tenses or verb forms that are the same for singular subjects as they are for plural subjects. In those contexts, an -ING word + plural noun combination is ambiguous.

Table 7.5 Tenses or Verb Forms in Which an -ING + Plural Noun Are Ambiguous

---

Past tense:

? Reporting violations caused the agency a lot of headaches.

Any tense in which *had* is the auxiliary verb:

? Reporting violations had caused the agency a lot of headaches.

? Reporting violations had been causing the agency a lot of headaches.

Future tenses:

? Reporting violations will cause the agency a lot of headaches.

? Reporting violations will be causing the agency a lot of headaches.

? Before long, reporting violations will have caused the agency a lot of headaches.

Constructions that include a modal verb:

? Reporting violations could cause the agency a lot of headaches.

? Reporting violations might cause the agency a lot of headaches.

Chapter 8

Punctuation and Capitalization

Introduction

In Global English, most of the guidelines for punctuation and capitalization are the same as the guidelines in conventional style guides such as *The Chicago Manual of Style*. However, conventional style guides don't take the needs of non-native speakers and translators into account. They have been written with the assumption that native speakers are the authors and the readers. Therefore, the guidelines in this chapter are intended to supplement (or, in a few cases, override) the standard rules of American English punctuation and capitalization.

Note: Strictly speaking, ampersands and equal signs are not punctuation marks. However, in this context, making the appropriate distinctions would interfere with clear communication.

Punctuation and Capitalization as Syntactic Cues

Like other syntactic cues, punctuation marks and capitalization help readers analyze and interpret sentences correctly. Notice how much more difficult it is to read the version of this paragraph that lacks these important syntactic cues:

- warning charge circuit timeout indicates that the charging period has exceeded 30 seconds the charge circuit is still active please inform your support representative if this device status indicator occurs immediate replacement is recommended
- Warning: Charge Circuit Timeout
  indicates that the charging period has exceeded 30 seconds. The charge circuit is still active. Please inform your support representative if this device status indicator occurs. Immediate replacement is recommended.

A sentence can be more difficult to read and comprehend even if only one punctuation mark or capital letter is missing. In the following sentence, even the conventional rules of punctuation call for a comma between the main clauses.

✗ The LET statement defines your macro variable and the macro variable references appear throughout the program statements.

✓ The LET statement defines your macro variable**,** and the macro variable references appear throughout the program statements.

Punctuation and Translation Memory

By now you know that in addition to encouraging the use of all types of syntactic cues, Global English aims to reduce unnecessary variation in all aspects of written language. Even slight variations can make the use of translation memory (TM) less effective. For example, consider the following scenario.

The following sentence was previously translated, so the English sentence and its translation are stored as a translation pair in a TM database:

▶ Click `Cancel` to return to the program and correct the error.

Later, a translator who is using the TM database encounters the following sentence:

▶ Click `Cancel` to return to the program**,** and correct the error.

The TM software flags this sentence as a fuzzy match. That is, it displays the previous sentence alongside the new one, highlights the comma in the new sentence, and displays the translation of the previous sentence. The translator compares the two sentences and determines whether the previous translation can also be used for the new sentence.

Even if the sentences are nearly an exact match, you pay the translator a certain amount per word for translating fuzzy matches. Thus, you incur a cost for every unnecessary variation. And you incur that cost not once, but for each language that your document is translated into.

Translators sometimes examine exact matches, too, if their TM database contains documents from different product areas or domains. (Sometimes the same sentence must be translated differently in different contexts.) However, comparing fuzzy matches takes the translator more time, and you pay more for that extra time and effort.

In addition, sometimes a difference in punctuation makes a subtle difference in meaning that the translator could easily overlook. In the following example, *correct* should be translated as an infinitive in the first sentence and as an imperative verb in the second:[1]

▶ Click `Cancel` to return to the program and **[to]** correct the error.

▶ Click `Cancel` to return to the program**,** and correct the error.

In other words, the first version implies that the single action of clicking `Cancel` has two consequences:

- The user is returned to the program.
- The error is corrected.

The second version makes it clear that correcting the error is a separate action that the user must perform.

Sometimes a slight difference in punctuation can even prevent TM software from recognizing two nearly identical sentences as fuzzy matches. In order to understand this, you need to understand how TM software divides texts into translation segments.

Most TM software uses the following punctuation marks as segment boundaries:

- periods
- question marks
- exclamation points
- semicolons
- colons

Translators can modify the segmentation rules, but these are the typical defaults. Since the following sentence contains only one of these segment boundaries (the period at the end), it constitutes a single translation segment:

▶ Every form has a current lock state—either locked or unlocked.

By contrast, the following sentence contains two translation segments, because both the colon and the period are segment boundaries.

▶ Every form has a current lock state: either locked or unlocked.

The difference in segmentation prevents TM software from identifying these two sentences as a fuzzy match, even though the only difference is that one has an em dash where the other has a colon.

---

[1] See guideline 8.3 for contexts in which commas can make a more significant difference in meaning.

As the following guidelines illustrate, there are many contexts in which most authors would consider either of two punctuation marks to be correct. An understanding of TM segmentation is often necessary in order to decide which punctuation mark to standardize on in those contexts.

8.1 Ampersands

Don't use an ampersand in place of the word *and*. This usage is a source of unnecessary variation. In addition, some non-native speakers are not familiar with this use of the ampersand.

- ✗ Use the Target Periods **&** Sources page to specify the source of data for each target period in the selected cycle.
- ✓ Use the Target Periods **and** Sources page to specify the source of data for each target period in the selected cycle.
- ✗ For more information, see Ch 14 **&** Ch 15.
- ✓ For more information, see chapters 14 **and** 15.

Of course, it is OK to use an ampersand to refer to itself:

- ✓ Special characters such as **&**, < , and > are not allowed.
- ✓ Note that **&** is replaced with the `&` character entity, and the entire URL is enclosed in single quotation marks.

8.2 Colons

In a sentence that introduces multiple possibilities, use an em dash rather than a colon:

- ✗ Every form has a current lock state: either locked or unlocked.
- ✓ Every form has a current lock state—either locked or unlocked.

In this context, most authors would find either a colon or an em dash acceptable. However, if you use a colon, then TM software typically treats the adjectives that follow the colon as a separate translation segment. (See "Punctuation and Translation Memory" on page 164.)

It is better for adjectives to be in the same translation segment as the nouns that they modify, because, in many languages, the adjectives must agree in number and gender with the nouns. For example, the French translations for *locked* and *unlocked* are slightly different depending on whether the noun that they modify is masculine or feminine:

X There are two types of files: locked and unlocked.

T Il y a deux sortes de fichiers: verrouill**és** et déverrouill**és**.

X There are two types of bank vaults: locked and unlocked.

T Il y a deux sortes de chambres fortes: verrouill**ées** et déverrouill**ées**.

In other cases, an adjective might have several translations, depending on the noun that it modifies. For example, *deep* might well be translated differently when it modifies *water* than when it modifies *discounts* or *neuroses*.

Alternatively, you could use a colon but repeat the noun that is being modified:

▶ There are two types of files: locked **files** and unlocked **files**.

This is the best approach if you are using machine-translation software. Otherwise, the additional words increase the cost of translation without providing any benefit.

8.3 Commas

8.3.1 Use commas to prevent misreading

Always use a comma if the comma will prevent misreading or will make the sentence structure clearer. Consider the following sentence:

X To develop an embryo needs the kind of lining that only the uterus has.

Many readers are initially confused by this sentence, because a sentence that begins with an infinitive (in this case, *To develop*) often introduces a procedure, as in the following example:

▶ To develop an embryo, follow these steps:

By inserting a comma following the infinitive, you can make it clear that the original sentence doesn't follow that pattern:

✓ To develop**,** an embryo needs the kind of lining that only the uterus has.

8.3.2 Use commas to separate main clauses

This is a standard rule of English punctuation. However, many authors don't follow it consistently. For example, they might fail to recognize main clauses in contexts like the following:

- clauses that have subjectless imperative verbs:

  ✓ Check the spelling of the target text expression in the %%GOTO statement**,** or add a label definition statement within the same macro definition.

- clauses that are framed by *either . . . or*:

  ✓ Either the list of parameters is not enclosed in parentheses**,** or the parameters in the list are not separated by commas.

8.3.3 Consider using a comma before *because*

All style guides agree that when a subordinate clause occurs at the beginning of a sentence, it should be separated from the main clause by a comma:

▸ **Because we are now able to confirm your order as soon as it is received**, your credit card charge will be processed immediately.

Most authors don't precede a *because* clause with a comma when it *follows* the main clause. But sometimes the comma is necessary in order to prevent misreading, misinterpretation, or confusion.

To make sure your meaning is clear, follow these guidelines:

- Use a comma before *because* when the clause that it introduces is not essential to your meaning. For example, if you omitted the *because* clause from the following sentence, you would have *You do not need to specify delimiters*. This is the main proposition of the sentence. The *because* clause merely adds an explanation.

  ✓ You do not need to specify delimiters, **because the blank and the comma are default delimiters**.

- Don't use a comma before *because* if the *because* clause is essential to the meaning. If you omitted the *because* clause from the first sentence in the following example, you would have *Don't use syntactic cues*. But the second sentence in the example makes it clear that that is not what the author is saying at all.

  ✓ Don't use syntactic cues **because they improve the output of machine-translation software**. Use them because they improve readability.

- Omit the comma if the *because* clause doesn't modify the entire preceding clause. In this example, the *because* clause explains why the interface is useful. In other words, it modifies *that is useful*, not the entire main clause.

  ✓ This example produces an interface that is useful **because customers can select values for columns instead of having to type values**.

 **Similarities between *Because* Clauses and Relative Clauses**

This guideline for the use of commas with *because* follows the same reasoning that is applied to restrictive and non-restrictive relative clauses.

- A restrictive relative clause provides essential information. No comma is used with a restrictive clause like the following:

  ✓ Butterfly taxa **that have a low risk of extinction** are listed in the Least Vulnerable column.

- A non-restrictive clause provides non-essential information. A non-restrictive relative clause begins with the relative pronoun *which*, *who*, *whom*, or *whose* and is preceded by a comma:

  ✓ Butterfly taxa, **which have a low risk of extinction**, are listed in the Least Vulnerable column.

As you can see, with relative clauses as well as *because* clauses, the meaning changes depending on whether the clause is restrictive or non-restrictive. The first sentence above implies that not all butterfly taxa have a low risk of extinction. The second sentence implies that they all do.

8.3.4 Consider using a comma before *such as*

In guideline 8.3.3 you saw that the presence or absence of a comma affects the meaning of *because* clauses and relative clauses. The same is true of phrases that begin with *such as*.

In the following sentence, the lack of a comma implies that webEIS documents are a type of Java applet:

X The JAR file format aggregates the files that are associated with Java applets **such as webEIS documents**.

That interpretation is incorrect. In the following revision, the comma makes it clear that webEIS documents are a type of file that is associated with Java applets:

✓ The JAR file format aggregates the files that are associated with Java applets**, such as webEIS documents**.

In other words, the comma makes it clear that *such as webEIS documents* modifies *the files that are associated with Java applets*, not just *Java applets*.

Here is an explanation of how to determine whether a comma is needed:

- If a *such as* phrase modifies only the closest preceding noun phrase, then don't precede *such as* with a comma.

  ✓ The primary component is a guide for specific property attributes **such as alignment attributes**. [*alignment attributes* are an example of *specific property attributes*]

  ✓ A browser enables a user to locate and view HTML documents from the World Wide Web or from local Webs **such as intranets**. [*intranets* are examples of *local Webs*]

  ✓ The analysis typically involves data-mining techniques **such as sequences and associations**. [*sequences and associations* are examples of *data-mining techniques*]

- If a *such as* phrase modifies more than just the closest preceding noun phrase (for example, a noun phrase plus a prepositional phrase or a noun phrase plus a relative clause), then precede *such as* with a comma.

  ✓ SAS/SHARE servers can now respond to requests from the host operating system**, such as shutdown commands and stop commands**. [*shutdown commands and stop commands* are not examples of *host operating systems* nor of *requests*; they are examples of *requests from the host operating system*]

  ✓ The report includes demographic information about the visitor population, site activity rates, and the relative demand for various areas of the site**, such as ad banners or links on a page**. [*ad banners or links on a page* are not examples of *the site* nor of *various areas*; they are examples of *various areas of the site*]

Follow this guideline even if the *such as* phrase occurs somewhere other than at the end of the sentence, as in this example:

✓ Most time-series analysis tasks assume that the observations are taken at equally spaced time intervals and that no time intervals are missing. Thus, a periodic series **such as transaction data** must be converted into periodic data before it is analyzed. [*transaction data* is an example of *a periodic series*, so no comma is needed]

8.4 Double Hyphens

Double hyphens are a relic from the days of typewriters. Don't use them in documents that are produced electronically. Aside from causing unnecessary inconsistency, double hyphens sometimes interfere with the use of language technologies such as controlled-authoring software.

Often you can use an em dash in place of a double hyphen. (See the next section for guidelines on the use of em dashes.) Some authoring tools can be set up to automatically convert double hyphens to em dashes. Alternatively, in Windows applications you can insert an em dash by holding down the Alt key and then pressing 0151 on the numeric keypad. On a Macintosh, hold down the Option key, and then press the hyphen key.

8.5 Em Dashes

In this section, guidelines 8.5.4–8.5.10 are aimed at eliminating unnecessary inconsistencies. There is nothing inherently wrong with using em dashes in the contexts that those guidelines describe. However, if you and your colleagues sometimes use em dashes and other times use parentheses to set off the same type of information, then your inconsistency will interfere with the efficient use of translation memory.

Guideline 8.5.11 lists some approved uses for em dashes.

8.5.1 Whenever possible, use a separate sentence instead

If a connector such as *that is*, *for example*, or *however* introduces a main clause or a subordinate clause that is followed by a main clause, then start a new sentence instead of using an em dash as a separator. Putting part of a long sentence into a separate sentence makes the use of translation memory and machine translation more efficient.

X A template is a single graphics element that you can resize, move, and delete—**that is, you cannot manipulate any part of the template separately, since it is only a placeholder for the actual graph**.

✓ A template is a single graphics element that you can resize, move, and delete. **That is, you cannot manipulate any part of the template separately, since it is only a placeholder for the actual graph**.

X If all applications store client address information in an address object and a change is needed in how this information is stored**—for example, to store the street address, the city and state, and the country in three separate fields instead of one—**then the change is easy to identify, make, and propagate.

✓ If all applications store client address information in an address object and a change is needed in how this information is stored, then the change is easy to identify, make, and propagate**. For example, you can easily store the street address, the city and state, and the country in three separate fields instead of one**.

8.5.2 Consider other ways of eliminating em dashes

X The fastest way to finish your configuration of the system**—and the safest way—**is to run the scripts that the `instructions.html` file refers to.

✓ The fastest **and safest way** to finish your configuration of the system is to run the scripts that the `instructions.html` file refers to.

X The order shown in the preceding list**—although not required—**is a good order in which to install these products because there are some dependencies.

✓ Because there are some dependencies, **you should** install these products in the order shown above.

8.5.3 Make sure the sentence would be grammatical if the em dash phrase were omitted

The first version of the following sentence is ungrammatical because the verb *are* doesn't agree in number with the subject *SLI region type*:

X For the DATA step interface, the SLI region type**—and hence, the SLICWTO, SLIREAD, and CICSID options—**are no longer supported.

✓ For the DATA step interface, the SLI region type**—and hence, the SLICWTO, SLIREAD, and CICSID options—**is no longer supported.

8.5.4 Don't use em dashes as a formatting device

In the following example, either a definition list or a hanging list would be appropriate:

✗ Analysis tab — contains functionality for calculating risk.

✓ Analysis tab
contains functionality for calculating risk.

✓ Analysis tab
    contains functionality for calculating risk.

In the next example, either a simple table or a block list should be used:

✗ Some common commands include M — moves a line of text, C — copies a line of text, D — deletes a line of text, I — inserts a line of text.

✓ Here are some common commands:

M	moves a line of text
C	copies a line of text
D	deletes a line of text
I	inserts a line of text

8.5.5 Don't use em dashes to set off cross-references

Use parentheses, not em dashes, to set off cross-references.

✗ Follow the same steps that you follow to upgrade a system in place—**see Upgrading in Place**—with the following caveat:

✓ Follow the same steps that you follow to upgrade a system in place (**see Upgrading in Place**), with the following caveat:

✗ When you configure the User Service—**as described in the next section**—you are asked to specify an administrator.

✓ When you configure the User Service (**as described in the next section**), you are asked to specify an administrator.

8.5.6 Don't use em dashes to set off definitions

Use parentheses, not em dashes, to set off definitions.

✗ The wizard uses the selected tables to generate a transformation template—**a process flow diagram that includes drop zones for metadata objects**.

✓ The wizard uses the selected tables to generate a transformation template (**a process flow diagram that includes drop zones for metadata objects**).

✗ The chart vertices—**the points where the statistical values intersect the spokes**—are based on the frequencies that are associated with the levels of a single numeric variable.

✓ The chart vertices (**the points where the statistical values intersect the spokes**) are based on the frequencies that are associated with the levels of a single numeric variable.

8.5.7 Don't use em dashes to set off examples

Use parentheses, not em dashes, to set off examples.

✗ If you expect to use a graph over and over again—**a company logo, for example**—then link to it so that you can make future changes in only one place.

✓ If you expect to use a graph over and over again **(a company logo, for example),** then link to it so that you can make future changes in only one place.

✗ Objects can be visual objects that you place on the frame—**for example, icons, push buttons, or check boxes**.

✓ Objects can be visual objects that you place on the frame (**for example, icons, push buttons, or check boxes**).

Keep in mind that examples are often introduced by *such as*:

✗ Some of the tabs—**such as the tabs for Notes, Support, and Process**—appear only if you have defined metadata for those tabs.

✓ Some of the tabs (**such as the tabs for Notes, Support, and Process**) appear only if you have defined metadata for those tabs.

8.5.8 Don't use em dashes to set off non-restrictive relative clauses

If possible, put non-restrictive relative clauses into separate sentences.

✗ Because stored process servers support MultiBridge connections—**which means that an object spawner can send requests to any one of a set of multi-user server processes**—it is possible to implement load balancing on a single host.

✓ Because stored process servers support MultiBridge connections, **an object spawner can send requests to any one of a set of multi-user server processes. Therefore,** it is possible to implement load balancing on a single host.

8.5.9 Don't use an em dash to introduce a complete sentence

Use a colon or a period, not an em dash, to introduce a complete sentence.

✗ If you are a Finance Adjuster, then you can use all the features of the Models workspace, with one exception—**you cannot create or edit unbalanced manual adjustments**.

✓ If you are a Finance Adjuster, then you can use all the features of the Models workspace, with one exception: **you cannot create or edit unbalanced manual adjustments**.

✗ You can store the autocall macros in as many libraries as you want—**but each time you call a macro, each library is searched sequentially until the macro is found**.

✓ You can store the autocall macros in as many libraries as you want. **However, each time you call a macro, each library is searched sequentially until the macro is found**.

✗ You can view the resulting PNG files in any browser—**neither Java nor ActiveX is required**.

✓ You can view the resulting PNG files in any browser. **Neither Java nor ActiveX is required.**

8.5.10 Don't use an em dash to introduce an -ING phrase

Use a comma, not an em dash, to introduce an -ING phrase.

✗ ActiveX draws each part of the step—**resulting in a somewhat different graph**.

✓ ActiveX draws each part of the step**, resulting in a somewhat different graph**.

See guideline 7.4, "Punctuate -ING phrases correctly," for additional examples.

8.5.11 Approved uses for em dashes

Emphasis

✓ A J2EE enterprise application uses other technologies—**in particular, Enterprise JavaBeans**—in addition to Java servlet technology.

✓ All processing is supported for tables—**if the engine that is defined in the metadata supports that processing**.

Enumerations

✓ In the following display, three MIME types—**ABC, ACGI, and AIP**—are counted as page views.

✓ The second request—**a GetMetadata method call**—retrieves the columns that you defined and verifies that the column was added.

Explanations that begin with *that is*

Use either em dashes or parentheses to set off explanations that begin with *that is*. This inconsistency is tolerated because, in this context, em dashes and parentheses have different rhetorical effects. Use em dashes for greater emphasis or for a greater break in thought. However, see also guideline 8.5.1, "Whenever possible, use a separate sentence instead."

✓ This problem sometimes occurs in a nested job—**that is, in a loop that iterates over another loop**.

✓ If this assumption is not true—**that is, if the regressor varies systematically with the error**—then ordinary regression produces inconsistent results.

Introducing adverb phrases

✓ The META2HTM macro is invoked twice—**once before the GRAPHGEN procedure in order to specify parameters for the procedure, and a second time after the procedure to close the HTML file that was created**.

Introducing noun phrases

✓ The TABLE OPTIONS property has a new sublist—**the APPEND sublist**.

✓ This example shows two LIBRARY statements—**one statement that uses default values, and one statement that specifies all of the LIBRARY statement options**.

Introducing prepositional phrases

✓ If the stratification variable is GENDER, then the flow loops twice**—for male and for female.**

✓ In the previous figure, information moves from the bottom up**—from ODD1**, **to a mapping step, to the Credit data table**.

8.6 Equal Signs

Don't use an equal sign in place of text. This usage causes unnecessary inconsistency and also causes problems for machine-translation software.

X Any non-blank character = turn off metadata checking, and blank = perform metadata checking.

✓ Any non-blank character **turns *off*** metadata checking, and a blank character **turns *on*** metadata checking.

X GRFSRC_ returns the type of graph that the user selected, where 1 = a clickable three-dimensional graph, and 2 = a standard GIF graph.

✓ GRFSRC_ returns a value that indicates which type of graph the user selected:

1 The user selected a clickable, three-dimensional graph.

2 The user selected a standard GIF graph.

However, it is OK to use equal signs to express option values in contexts like the following:

✓ When MISSING=PAIRWISE, the calculation of the diagonal element is based on the number of non-missing observations.

8.7 Hyphens

8.7.1 Consider hyphenating noun phrases

Hyphens can make noun phrases easier for readers to interpret correctly, and they are sometimes essential for translation.

X The value in register 1 must be saved in the **argument string word** of the return value.

✓ The value in register 1 must be saved in the **argument-string word** of the return value.

X **Site specific** configuration files are typically set up by your system administrator in order to make the best use of your **operating environment** resources.

✓ **Site-specific** configuration files are typically set up by your system administrator in order to make the best use of your **operating-environment** resources.

Some noun phrases—especially those that begin with the adjective *more*—have different meanings depending on whether a hyphen is used, as in these examples:

X **More general** models might provide fewer options than **more specific** models.

✓ **More-general** models might provide fewer options than **more-specific** models.

For additional details about clarifying the meaning of noun phrases, see guideline 3.7.2, "Consider revising noun phrases."

8.7.2 Use hyphens consistently in the noun and adjective forms of phrasal verbs

Multi-word verbs such as *back up* and *drop out* are usually spelled as separate words when they are used as verbs. When they are used as nouns, the spelling of these words is often inconsistent: Sometimes they are spelled as one word, and other times they are hyphenated.

To eliminate such inconsistencies, follow these guidelines:

- If the word is spelled as one word (no hyphen) when it is used as a noun, then spell the adjectival form like the noun, as in these examples:

set up	verb	Google announced plans to **set up** back-office operations in a special economic zone in India.
setup	noun	We estimated that we needed $25,000 for office **setup** and initial networking.
setup	adjective	Please note that some advanced **setup** issues are not discussed here.

lock down	verb	Using IPSec policies to **lock down** a server provides greater flexibility.
lockdown	noun	Desmond triggered a **lockdown** by briefly shorting two power-supply wires.
lockdown	adjective	When the **lockdown** procedure terminates, the power in the hatch is automatically restored.

work around	verb	There are several ways to **work around** the Outlook security update.
workaround	noun	The **workaround** described in Usage Note 8132 does not need to be implemented with the version 3.3 drivers.
workaround	adjective	RMI has incorporated the **workaround** designs into a software update called the BoisenBerry Enhanced Edition.

- If the noun form is hyphenated, then hyphenate the adjectival form as well:

start up	verb	My PC takes a long time to **start up** or reboot.
start-up	noun	If the **start-up** doesn't have a business plan or is unwilling to show it to you, you probably don't want to work there.
start-up	adjective	Raising **start-up** capital is one of the biggest challenges that entrepreneurs face.

give away	verb	For tax reasons, the foundation must **give away** every nickel that was contributed during the previous year.
give-away	noun	The organizations held a **give-away** of the new Mindprint TRIO USB.
give-away	adjective	In honor of its twentieth anniversary, the company is hosting a **give-away** contest.

8.8 Parentheses

8.8.1 Make sure readers can understand what parentheses are intended to indicate

Use parentheses to indicate synonyms or equivalents

Use parentheses to indicate that two terms are synonyms or that two values are equivalent, as in these examples:

✓ You can also use the graphical user interface **(GUI)** to issue commands.

✓ Ignoring the curvature of the earth over a distance of 100 miles **(161 kilometers)** results in approximately 14 feet **(4.3 meters)** of error.

Don't use parentheses to indicate alternatives

As noted above, you can use parentheses to indicate that two terms are synonyms or that two values are equivalent. But don't use parentheses to indicate that one term is an alternative to another.

The following sentence leads readers to believe that E-Plus is the new name of the company that was formerly known as Mobifunk. In reality, they are different companies.

✗ With cellular service provided by E-Plus **(Mobifunk)**, you can travel securely and conveniently with your own German cellular phone.

In the following revision, the use of *either . . . or* makes it clear that the two companies are distinct entities:

✓ With cellular service provided by **either** E-Plus **or** Mobifunk, you can travel securely and conveniently with your own German cellular phone.

Don't use an open parenthesis followed by *or*

Even if you use parentheses to indicate that two terms are synonyms (rather than alternatives), readers and translators won't be sure what you mean if you precede the synonym with *or*. This usage is not allowed because *or* can imply either that the terms are synonyms or that they are alternatives. Context often is not sufficient to make the meaning clear.

Here is an example in which a parenthesis followed by *or* introduces ambiguity and confusion for translators and for other readers:

✗ You can use an Explorer **(or DIR)** window to perform several common tasks:

The relationship between *Explorer* and *DIR* is unclear. This sentence provides a nice segue to the next topic:

Make semantic relationships crystal clear

Even if two terms are synonyms, you can't always rely solely on parentheses to make that semantic relationship clear. If the terms are very dissimilar, then you might need to make their relationship explicit. For example, omitting *or* from the previous example doesn't resolve the confusion in readers' minds:

✗ You can use an Explorer **(DIR)** window to perform several common tasks:

The author should explain what *DIR* is in a separate sentence—perhaps in a footnote:

✓ You can use an Explorer window[1] to perform several common tasks: . . . [1] In the z/OS operating environment, you can use the DIR window to perform these tasks.

Here is another example that baffles most readers:

✗ These options cause each value in the transaction data set to generate a new result set **(open cursor)** from the table.

The terms *result set* and *open cursor* are so dissimilar that you must explain their relationship before using them in this manner.[2]

8.8.2 Make parenthetical information grammatically independent

Parenthetical information must always be grammatically independent of the rest of the sentence. Since it is ungrammatical to pair a singular noun with a plural verb, the parentheses must be removed from these example sentences:

✗ The table **(and all rows)** are dropped when the connection is closed.

✓ The table and all rows are dropped when the connection is closed.

✗ Test the server **(and clients)** that enable your users to schedule sets of jobs.

✓ Test the server and clients that enable your users to schedule sets of jobs.

8.8.3 Whenever possible, put parenthetical information in a separate sentence

✗ All sessions have an associated timeout value **(the default is 15 minutes)**.

✓ All sessions have an associated timeout value. **The default is 15 minutes.**

✗ The `Arrange` item aligns the icons for the data objects **(in multiple rows, if needed)**.

[2] A corrected version of the sentence is not presented here because the topic in which the sentence was found did not include an explanation.

✓ The `Arrange` item aligns the icons for the data objects. **If necessary, the icons are arranged in multiple rows.**

✗ The entries in the list might be descriptive, such as Server, or they might be actual host names **(in custom plans)**.

✓ The entries in the list might be descriptive, such as Server. **In custom plans,** the entries might be actual host names.

✗ Use this window to add additional types of SAS servers **(consisting of a logical server and a corresponding physical server)** to the main server definition.

✓ Use this window to add additional types of SAS servers to the main server definition. **Each SAS server consists of a logical server and a corresponding physical server.**

✗ A Leaf members formula determines the values for all locations that are designated by the formula-bearing member **(which must be a leaf member)**, as well as the values of leaf members for all other dimensions.

✓ A Leaf members formula determines the values for all locations that are designated by the formula-bearing member, as well as the values of leaf members for all other dimensions. **The formula-bearing member must be a leaf member.**

8.8.4 Eliminate unnecessary parentheses

If parentheses don't make a sentence more concise or more readable, and if the information that they contain is not clearly parenthetical, then they are simply intrusive.

✗ If you specify an output data set, then all the variables from the input address data set are copied to the output data set **(and the new variables are added)**.

✓ If you specify an output data set, then all the variables from the input address data set are copied to the output data set**, and the new variables are added**.

✗ Prompting enables you to set more options than you can set when you connect directly to the provider **(instead of using OLE DB Services)**.

✓ Prompting enables you to set more options than you can set when you connect directly to the provider **instead of using OLE DB Services**.

✗ Use the Class Settings window to specify whether a class **(and its objects)** can use the Properties window, a custom attributes window, or both, for setting properties.

✓ Use the Class Settings window to specify whether a class **and its objects** can use the Properties window, a custom attributes window, or both, for setting properties.

8.8.5 Eliminate parenthetical comments that impede readability

Parentheses often enable an author to add information to a sentence concisely without making the sentence structure excessively complicated. For example, both versions of the following sentence are correct, but the second is more concise, easier to read, and hence easier and less expensive to translate:

✓ The FORMULA= option is typically used to create formula variables in the output data view. If you also specify the OUTDATA= option, then you can also create derived variables in the output data set. [34 words]

✓+ The FORMULA= option is typically used to create formula variables in the output data view **(and to create derived variables in the output data set, if OUTDATA= is specified)**. [29 words]

But in the following example, parentheses were used to cram so much information into the sentence that the sentence is difficult to comprehend and translate:

✗ The program sets the values of items in the view descriptor **(that is, only the selected items in the same record)** to blanks and removes the lowest-level data record from the database.

Untangle the ideas and put them into separate sentences instead:

✓ **When the selected items are in the same record,** the program sets the values of those items in the view descriptor to blanks. The program also removes the lowest-level data record from the database.

8.8.6 Don't use *(s)* to form plural nouns

In English, *(s)* is commonly used to indicate that the reader should interpret a noun as either singular or plural:

✗ Check the **error(s)** in the log file.

However, in most languages, you cannot make a noun plural simply by adding a suffix to it. Therefore, the noun that ends in *(s)* often is not directly translatable. The *(s)* forces translators to make a choice between the singular and plural interpretation or to revise the sentence substantially. When a document is being translated into several languages, each translator must spend extra time deciding what to do about *(s)*.

You can eliminate *(s)* constructions by using the revision strategies described below.

More about *(s)* and Translation

In some languages and contexts, *(s)* actually *can* be translated directly. For example, you can do it with some German feminine nouns if they are the subject or direct object of the sentence, because the plural is formed simply by adding *-en*:

▶ Are you sure you want to delete the **file(s)**?

T Möchten Sie die **Datei(en)** wirklich löschen?

You can also do it in French sometimes. However, French articles, adjectives, participles, and pronouns must agree in gender and number with the nouns that they modify or refer to. Thus, a single instance of *(s)* in English can proliferate to many instances in the French translation:

▶ The **file(s)** referenced in the stored process must be created and maintained by someone who has Write access to the files' locations on the system.

T Le**(s)** fichier**(s)** référencé**(s)** dans l'application stockée doi**t(vent)** être créé**(s)** et actualisé**(s)** par un utilisateur ayant un accès en écriture à l'emplacement des fichiers sur le système.

Translators have various strategies for dealing with *(s)*, but many of those strategies (like the one above) are awkward and unpleasant. You wouldn't want to see comparable strategies used in English, so why impose them on readers of your translated documentation? Use the following revision strategies to eliminate *(s)* instead.

Use *each, every,* or *any of*

✗ Return the **column(s)** to their original **position(s)**:

✓ Return **each column** to its original **position**:

✗ The **file(s)** that are referenced in the stored process must be created and maintained by someone who has Write access to the files' locations on the system.

✓ **Every file** that is referenced in the stored process must be created and maintained by someone who has Write access to the file's location on the system.

✗ When the EFI encounters the specified **character(s)** in the external file, it recognizes them as special missing values.

✓ When the EFI encounters **any of** the specified **characters** in the external file, it recognizes them as special missing values.

Use *one or more*

X You will need your installation **kit(s)**, which contains the CDs or DVDs that you will use during the installation.[3]

✓ You will need **one or more** installation **kits**, which contain the CDs or DVDs that you will use during the installation.

Rearrange the sentence

X Check the **error(s)** in the log file.

✓ Check the log file for **errors**.

Use just the plural noun

Consider using just the plural noun. The loss of meaning usually is not significant.

X Check the specified system error **table(s)**, and correct the errors.

✓ Check the specified system error **tables**, and correct the errors.

This revision strategy is often the only one that is suitable for user-interface text. In the following example, a singular noun would be misleading:

X The Create **a** Job **Object(s)** for Scheduling dialog box contains the following fields:[4]

⊘ The Create a Job **Object** for Scheduling dialog box contains the following fields:

The "one or more" approach that is used below is unnecessarily precise, considering that few users are going to pay significant attention to the title of the dialog box:

⊘ The Create **One or More** Job **Objects** for Scheduling dialog box contains the following fields:

The world will not end if you just use the plural:

✓ The Create Job **Objects** for Scheduling dialog box contains the following fields:

---

[3] Note the conflict between the plural interpretation of *kit(s)* and the verb *contains*. This usage is a violation of guideline 8.8.2, "Make parenthetical information grammatically independent."

[4] The conflict between the singular article *a* and the plural interpretation of *Object(s)* is another violation of guideline 8.8.2. This conflict is far more disturbing than the slight loss of meaning that occurs when you make the noun plural.

Use just the singular noun

In user-interface text, if the noun can be either a mass noun or a countable noun, then consider using the singular:

X On the Confirm **Deletion(s)** page, answer the following prompts:

✓ On the Confirm **Deletion** page, answer the following prompts:

Use *<singular noun> or <plural noun>*

Sometimes you can use *<singular noun> or <plural noun>* instead of *(s)*, as in these examples:

X Select the **year(s)** for which the report will be generated.

✓ Select the **year or years** for which the report will be generated.

X If you have several SID files, you might want to rename the SID file so that it is clear which **machine(s)** the SID file is for.

✓ If you have several SID files, you might want to rename the SID file so that it is clear which **machine or machines** the SID file is for.

However, often this is not the best revision strategy. Always consider using one of the other revision strategies instead.

X Select the **item(s)** that you want to remove.

✓ Select the **item or items** that you want to remove.

✓+ Select **the items** that you want to remove.

X Use the administration tool to create the appropriate group **folder(s)** for the content.

✓ Use the administration tool to create the appropriate group **folder or folders** for the content.

✓+ Use the administration tool to create **one or more** group **folders** for the content.

(s) in Software Messages or Labels

In software messages or user-interface labels, *(s)* can be a huge problem. Unfortunately, it's a problem that translators encounter frequently. Consider the following example, in which {0} represents a number (possibly 0, but usually 1 or more) that is inserted dynamically when the actual message is generated:

- ▶ {0} scenario(s) were deleted.

In German, the translation is different if {0} is replaced by 1 than it is if {0} is replaced by 0 or by any other number:

- T **1** Szenario **ist** gelöscht worden.
- T **0** Szenari**en sind** gelöscht worden.
- T **2** Szenari**en sind** gelöscht worden.

In Russian and other Slavic languages, the translations are different depending on whether there are no scenarios, 1 scenario, 2-4 scenarios, or 5 or more scenarios:

- T **0** сценари**и** был**и** удален**ы**.
- T **1** сценари**й** был удален.
- T **2** сценари**я** был**и** удален**ы**.
- T **5** сценари**ев** был**и** удален**ы**.

The German translators would probably have to choose a translation like this:

- T {0} Szenari**o**/Szenari**en ist/sind** gelöscht worden.

The Russian translator would probably have to choose a translation like this:

- T {0} сценари**и**/сценари**й**/сценари**я**/сценари**ев** был/был**и** удален/удален**ы**.

How readable is the English equivalent of the Russian translation?

- ▶ {0} scenarios/scenario/scenarios/scenarios was/were deleted/deleted.

(*continued*)

Instead of using *file(s)*, the software developer should implement separate software messages to account for differences in singulars and plurals. If you separate the number of scenarios from the rest of the sentence structure, then even for Russian you might need only two separate messages:

▶ No scenarios were deleted.

T Ни одного сценария не удалено.

▶ Some scenarios have been deleted. Number of deleted scenarios: {0}

T Некоторые сценарии были удалены. Количество удаленных сценариев: {0}

8.8.7 Approved uses for parentheses

Acronyms

Use parentheses for including an acronym or some other short form of a term after the full form of a term. Also use parentheses for including the full form of a term after an acronym.

✓ A SAS package file **(SPK file)** is a container file.

✓ Instead of creating an entry for each user, you can create an LDIF **(Lightweight Directory Interchange Format)** file.

Concise Alternatives to Separate, Complete Sentences

Use parentheses to enclose concise phrases that you would otherwise have to put in separate sentences.

✓ A value of 1 **(the default)** specifies to compare only identities with which an ExternalIdentity object is associated.

✓ **(optional)** Add a background image to the banner.

Cross-references

Use parentheses to enclose cross-references.

✓ After verifying that users can connect to the metadata server **(as described in Connecting to SAS Servers)**, make sure that they can access the data sources as well.

For consistency, when a cross-reference occurs at the end of a sentence, always place it in a separate sentence.

✗ The user login functions as an inbound login **(see Important Security Terms)**.

✓ The user login functions as an inbound login. **(See Important Security Terms.)**

✗ A LAYOUT LATTICE enables you to display a sidebar between a header and an axis **(see the first figure for LAYOUT LATTICE)**.

✓ A LAYOUT LATTICE enables you to display a sidebar between a header and an axis. **(See the first figure for LAYOUT LATTICE.)**

Definitions

Use parentheses to enclose definitions of terms.

✓ A J2EE Web application is delivered as a WAR file **(an aggregate file that contains all of the files that make up the application)**.

Examples

If you need to set off examples from the rest of the sentence, then use parentheses rather than em dashes.

✓ Members of the Portal Admin group **(for example, the SAS Web Administrator)** are automatically designated as group content administrators for all groups.

✓ A class describes an object's characteristics **(such as attributes or instance variables)**, as well as the operations that the object can perform.

✓ Area features represent two-dimensional entities such as geographic areas **(countries, states, and so on)** or floor plans for buildings.

✓ A Grid layout is commonly used to position a text entry **(title, footnote, or legend)** outside the graph area.

Explanations

Use parentheses to enclose most types of explanations.

✓ A left outer join lists matching rows and rows from the left-hand table **(the first table listed in the FROM clause)** that do not match any row in the right-hand table **(the second table listed in the FROM clause)**.

✓ Filters are evaluated based on the data type **(character or numeric)** of the selected data item and the locale that is currently active for the browser.

✓ A copy of the log is stored in the My Documents folder **(on Windows)** or in your home directory **(on UNIX)**.

✓ An optional _WAIT parameter can be used to specify a maximum wait time **(in seconds)** for any sessions to expire.

Explanations that begin with *that is*

Use either em dashes or parentheses to set off explanations that begin with *that is*. This inconsistency is tolerated because, in this context, em dashes and parentheses have different rhetorical effects. Use em dashes for greater emphasis or for a greater break in thought. However, see also guideline 8.8.3, "Whenever possible, put parenthetical information in a separate sentence."

✓ An object cannot receive a _refresh method unless it is completely initialized **(that is, unless the object's _init and _postInit methods have run)**.

Special characters or punctuation marks

Technical documentation sometimes contains references to special characters or punctuation marks. The characters are often enclosed in parentheses, as in these examples:

✓ Special characters generally are not allowed. However, in aliases you can use the dollar sign **($)**, the pound sign **(#)**, and the at sign **(@)**.

✓ Numeric missing values are represented by a single decimal point **(.)**.

It probably is not necessary to show readers what periods and commas look like—although periods are referred to as full stops in some varieties of English. However, for other characters and punctuation marks, this convention is appropriate for a global audience. Non-native speakers might not recognize the names of characters such as tildes (~) and asterisks (*). In addition, some of these marks are not the same for all languages. For example, the appearance and usage of quotation marks varies significantly around the world, and in Danish, the English division sign (÷) is used to indicate subtraction.

Synonyms or equivalents

✓ You can use the customer demo **(custdemo)** account to test the authentication.

✓ Users can drill down to a map of Wake County **(MAPS.WAKE.TRACT)** by selecting `Wake County` and running the DRILL action.

8.9 Quotation Marks

8.9.1 Don't use quotation marks to represent inches or feet

Many readers in countries that use the metric system do not understand the meaning of quotation marks that are used for this purpose. Also consider including the metric equivalents of English measurements.

X By default, the shadow extends 1/16" to the right and 1/16" below the legend.

✓ By default, the shadow extends 1/16th **of an inch** to the right and 1/16th **of an inch** below the legend.

X Ignoring the curvature of the earth over a distance of 100 miles results in approximately 14' of error.

✓ Ignoring the curvature of the earth over a distance of 100 miles (**161 kilometers**) results in approximately 14 **feet (4.3 meters)** of error.

8.9.2 Don't use quotation marks for metaphors

Don't use quotation marks to indicate that a term is being used in a metaphorical sense. Non-native speakers might not understand such terms, and it might be difficult for translators to express the metaphor in other languages. Replace the metaphor with literal language.

X If you invoke the Import wizard on a PC that contains an ATI XCEL128 video card, the PC might **"freeze."**

✓ If you invoke the Import wizard on a PC that contains an ATI XCEL128 video card, the PC might **stop responding to input**.

8.9.3 Don't use quotation marks for technical terms

Don't enclose a term in quotation marks merely because it is being used in a specialized or technical sense.

✗ The Foundation repository is a **"master"** repository on which other repositories rely for shared information.

✓ The Foundation repository is a **master** repository on which other repositories rely for shared information.

8.10 Semicolons

8.10.1 Don't use semicolons to separate clauses

Some authors use semicolons extensively as clause separators, but most do not. Therefore, this use of semicolons contributes to unnecessary variation.

Because semicolons are not followed by capital letters, they blend in with the surrounding text. They can make already-long sentences seem exceptionally long and intimidating, as in this example:

✗ Calendar 1 has a holiday on Sunday; on Monday and Tuesday work is done from 7:00 a.m. to 11:00 a.m. and from 12:00 noon to 4:00 p.m.; on Friday work is done from 12:00 (midnight) to 8:00 a.m.; on Saturday work is done from 7:00 a.m. to 11:00 a.m.; **on** other days work is done from 9:00 a.m. to 5:00 p.m., as defined by the default calendar.

In this case, most technical writers would put the information in an unordered list even if they were not writing for a global audience:

✓ Calendar 1 has the following characteristics:

- Sunday is a holiday.
- On Monday and Tuesday, work is done from 7:00 a.m. to 11:00 a.m. and from 12:00 noon to 4:00 p.m.
- On Friday, work is done from 12:00 (midnight) to 8:00 a.m.
- On Saturday, work is done from 7:00 a.m. to 11:00 a.m.
- On other days, work is done from 9:00 a.m. to 5:00 p.m., as defined by the default calendar.

In technical documentation there is seldom a significant difference in the rhetorical effect of semicolons and periods when they are used as clause separators. In particular, when the second clause starts with a logical connector such as *that is*, *for example*, or *however*, the difference in the rhetorical effect is negligible:

✗ Because the user ID is a trusted ID, servers such as the OLAP server, object spawner, and mid-tier applications can use this ID to impersonate authenticated clients on the metadata server**; that is,** the servers can communicate with the metadata server on behalf of the clients.

✓ Because the user ID is a trusted ID, servers such as the OLAP server, object spawner, and mid-tier applications can use this ID to impersonate authenticated clients on the metadata server**. That is,** the servers can communicate with the metadata server on behalf of the clients.

✗ Objects can also be nonvisual objects that manage the application behind the scenes**; for example,** an object that enables you to interact with data might not have a visual representation, but it can still enable you to access variables, add data, or delete data.

✓ Objects can also be nonvisual objects that manage the application behind the scenes**. For example,** an object that enables you to interact with data might not have a visual representation, but it can still enable you to access variables, add data, or delete data.

Even when no logical connector is present, the rhetorical difference between a semicolon and a period is usually insignificant:

✗ Only activities that have actual start times are assumed to have started**; all** other activities have an implicit start time that is greater than or equal to TIMENOW.

✓ Only activities that have actual start times are assumed to have started**. All** other activities have an implicit start time that is greater than or equal to TIMENOW.

8.10.2 When necessary, use semicolons to separate items in a series

Although you should not use semicolons as clause separators, Global English doesn't impose any other extraordinary restrictions on the use of semicolons. For example, when one or more items in a series includes a comma, you can use semicolons to separate the items:

✓ Packages can contain SAS files (SAS catalogs**;** SAS data sets**;** various types of SAS databases, including cubes**;** and SAS SQL views), binary files, HTML files, reference strings, text files, and viewer files.

In this context, semicolons are useful syntactic cues that help prevent misreading.

Sometimes semicolons are also useful for clarifying which items in a series a prepositional phrase or relative clause is modifying. See guideline 4.3, "Clarify what each relative clause is modifying," for an example.

8.11 Slash

In general, avoid using slashes to join two or more words. This construction can be difficult to translate. It doesn't have the same meaning in all contexts, and translators sometimes cannot determine what you mean.

For example, in the following sentences, it is not clear whether the terms that are joined by slashes are synonyms or whether they are alternatives to each other:

- ✗ The following example shows how to move a folder that contains **publication/subscription** content:
- ✗ If any entry type is 'ALL', then all **entries/variables** will be displayed regardless of type.

In the next example, it is not clear whether COM/DCOM represents a single entity or what the relationship is between the two acronyms:

- ✗ This application programming interface (API) provides access to the server from a variety of programming environments, including Java, **COM/DCOM**, and SAS.

On the other hand, some terms that are joined by slashes are well established in particular subject areas and should not be changed. For example, it would be absurd to change the term *client/server* in software documentation:

- ▶ A sample **client/server** XML log is shown below.

In addition, it would be silly to make a change in some contexts:

- ▶ A **Y/N** switch indicates whether to allow users to use the USER= and PASSWD= options.

The following subsections present strategies for dealing with joined terms that *do* need to be changed or explained.

8.11.1 Submit unavoidable joined terms to your localization coordinator

If a joined term in a document that will be translated is uncommon or is used in very specific technical contexts, then submit the term to the localization coordinator before localization begins. The localization coordinator can then ask you for explanations of any joined terms that she or he has not encountered before.

Here are some examples of joined terms that a localization coordinator would probably want to know about in advance:

- The Model Viewer graphically displays the interdependence of the most significant risk factors and their effects on **profit/loss**.
- You can specify filtering criteria in the form of **name/value pairs**.
- Interfaces help implement **model/view communication** by establishing a type of relationship between the model and the viewer.

It is especially important to communicate with the localization coordinator about any joined terms that are used in a software user interface. Software localization typically begins before the documentation is localized. Therefore, when a localization coordinator encounters user-interface labels like the ones below, there is no documentation to refer to for explanations:

- Allow horizontal **Grow/Shrink** but not vertical
- Offer **group/subgroup**
- **Value/CAMs**

8.11.2 Use *or* instead

If the joined terms are alternatives to each other, you can often separate the terms with *or*. However, in many contexts, if you merely replace the slash with *or*, you will introduce an ambiguity. You might need to use one of the revision strategies that are discussed in guideline 4.6, "Clarify ambiguous modification in conjoined noun phrases."

In the revision of the following example, it is not clear that *page* modifies *size*:

✗ Choosing **a page/buffer size** that is larger than the default can speed up execution time.

⊘ Choosing **a page or buffer size** that is larger than the default can speed up execution time.

Only the second of the following possible interpretations is correct:

⊘ Choosing **a page or <u>a</u> buffer size** that is larger than the default can speed up execution time.

✓ Choosing **a page <u>size</u> or <u>a</u> buffer size** that is larger than the default can speed up execution time.

8.11.3 Separate the joined terms with *and* or with a comma

If the joined terms are alternatives to each other, you can usually just separate them with a comma. In the following example, the COM and DCOM acronyms represent separate but related standards for communication among software components:

✗ This application programming interface provides access to the server from a variety of programming environments, including Java, **COM/DCOM**, and SAS.

✓ This application programming interface provides access to the server from a variety of programming environments, including Java, **COM, DCOM**, and SAS.

If these two alternatives occur at the end of a sentence, then of course they also need to be separated by *and*:

✓ This application programming interface provides access to the server from a variety of programming environments, including Java, SAS, **COM, and DCOM**.

8.11.4 Eliminate unnecessary synonyms

If the joined terms are exact synonyms, consider standardizing on one term. If necessary, find an appropriate place in your document to acknowledge both terms but to explain that you are using only one of them.

If the joined terms are *near* synonyms, or if most of what you say about one of the terms in part of your document applies to the other term as well, then try to find a term that encompasses both of them, as in this example:

✗ If you are creating the account on Windows, grant the `Log on as a batch job` **permission/policy** for the account on each Windows machine in the cluster.

✓ If you are creating the account on Windows, grant the `Log on as a batch job` **authorization** for the account on each Windows machine in the cluster.

8.12 Slash used in *and/or*

The *and/or* construction is a special case. Unlike other joined terms, *and/or* is almost always clear and concise. It doesn't pose a problem for non-native speakers or other readers, and it doesn't pose a problem for translation into languages such as German and French.

Consult your localization coordinator before you decide to eliminate *and/or* from documents that will be translated. If you decide to eliminate it, then consider using one of the following revision strategies.

8.12.1 Use *a, b, or both*

If *and/or* joins two elements at the end of a sentence, consider using *or both* instead:

- ✗ You can avoid this problem by selectively ordering the plot statements **and/or** by using transparency on the individual plots.
- ✓ You can avoid this problem by selectively ordering the plot statements, by using transparency on the individual plots, **or both**.
- ✗ To override the sequential order for filling cells, you can use the COLUMN= **and/or** ROW= options, which give you complete control over the fill order.
- ✓ To override the sequential order for filling cells, you can use the COLUMN= option, the ROW= option, **or both**. These options give you complete control over the fill order.

8.12.2 Use *any of the following* or *one or more of the following*

If *and/or* joins more than two elements at the end of a sentence, consider using *any of the following* or *one or more of the following*:

- ✗ You can add context-sensitive Help for components, for status-line messages, **and/or** for tool tips.
- ✓ You can add context-sensitive Help for **any of the following** items: components, status-line messages, and tool tips.
- ✗ A simple portlet displays text, data, **and/or** images.
- ✓ A simple portlet displays **one or more of the following** types of content:
  - text
  - data
  - images

8.12.3 Use only *or* or only *and*

If *and/or* joins elements that are not at the end of a sentence, then consider replacing it with just *or* or *and*. There is some loss of meaning or precision, but the loss is seldom significant:

✗ If you have entered your **user ID and/or password** incorrectly, you have two more opportunities to enter the correct information.

✓ If you have entered your **user ID or password** incorrectly, you have two more opportunities to enter the correct information.

✗ the Add Users **and/or** Groups window

✓ the Add Users **or** Groups window

✗ Select `System` to review **and/or** select a font from available system fonts.

✓ Select `System` to review the available system fonts **and** to select a font.

⚠ Be Careful Not to Distort Your Meaning

In the next example, changing *and/or* to *or* would imply that the system administrator has not performed any of the three actions. That interpretation is incorrect.

✗ The system administrator has not installed, configured, **and/or** started the WebDoc application.

⊘ The system administrator has not installed, configured, **or** started the WebDoc application.

This sentence illustrates how difficult it can be to find a suitable alternative for an *and/or* construction.

The following revision is unacceptable because of the sexist pronoun *he*:

⊘ The system administrator has not installed or configured the WebDoc application, or he has not started it.

The next revision is OK except that it uses passive voice:

? The WebDoc application has not been installed or configured, or it has not been started.

This last revision is perhaps the most suitable for translation even though it is eight words longer than the original sentence:

✓ Either the system administrator has not installed or configured the WebDoc application, or the system administrator has not started the application.

8.12.4 Revise more substantially

As always, search for alternatives if the revision strategies discussed in this book are not optimal for a particular context. For example, sometimes the elements that *and/or* joins are not at the end of a sentence, but you can move them to the end by dividing or rearranging the sentence:

✗ A simple portlet displays text, data, **and/or** images, with no interactive capabilities.

✓ A simple portlet displays **one or more of the following** types of content: text, data, images. **It does not provide any** interactive capabilities.

Here is another possible revision of that sentence:

✓+ A simple portlet displays text, data, images, **or any combination of the three. It does not provide any** interactive capabilities.

In the revision of the next example, parentheses were used to enclose the alternatives that were being joined by *and/or*:

✗ You can specify up to four SIDEBAR blocks in a LAYOUT LATTICE—one for each of the top, bottom, left, **and/or** right sidebar positions.

✓ You can specify up to four SIDEBAR blocks in a LAYOUT LATTICE—one for each of the sidebar positions (top, bottom, left, right).

8.13 Capitalization

The Importance of Consistent Capitalization

Capitalization is an important syntactic cue because it helps translators and other readers recognize that a term is a proper noun (name) rather than a common noun. Common nouns are virtually always translated. Names of companies and products are not translated.

Other proper nouns, such as names of major software components, might or might not be translated.[5] Nevertheless, it is important to be consistent about whether you capitalize a

[5] Different cultures, and different audiences within those cultures, have different attitudes toward terms that are left in English. In one software document in which *JDBC Table Viewer* was used, the term was left in English in the German version of the document. In the French version of the document, the term was translated.

particular term or not. Translators should be able to rely on English capitalization to help them decide whether to translate a term or not.

Here is an example of what can happen when you are inconsistent about capitalization. Suppose that in one part of a document, the term *response table base name* is not capitalized. A German translator translates it because it appears to be a common noun.

▶ Specify a **response table base name**.

T Geben Sie ein **Basisname für die Response-Tabelle** an.

Elsewhere—in a related document or in a software message—the same term is capitalized:

▶ A **Response Table Base Name** must be specified.

Now the translator has to determine whether the term is a proper noun or not—a time-consuming process that can require the involvement of several other individuals. In the worst-case scenario, the capitalized term is left in English in the second sentence. After all, there is no guarantee that the same translator will translate all texts or software in which the term occurs, nor that teams of translators will always be able to collaborate closely when faced with tight deadlines.

T Es muss ein **Response Table Base Name** angegeben werden.

Later, a user of the German version of the software needs additional information about this *Response Table Base Name*. But the detailed information is in a topic in which the term was translated because the English term was in lowercase. The German user somehow has to know to look for information about the *Basisname für die Response-Tabelle* instead of about the *Response Table Base Name*. How likely is the German user to find the information?

Perhaps the user will be able to get the necessary information from technical support—at additional expense to the software company, of course. But in the future, if he has a choice, he might choose to avoid such problems by buying software that was initially developed in German rather than a product that was translated from some other language.

In summary, inconsistent capitalization can have the following consequences:

- Translators are less productive because they have to research capitalization issues.
- Subject-matter experts and others must spend time helping to resolve such issues.
- Inconsistencies in English capitalization lead to inconsistencies in translations.

- Users or readers of translated software or documentation are frustrated or confused by inconsistencies. The inconsistencies force them to contact technical support for assistance.
- Users' frustrations influence their future choices of products.

The guidelines in this section will help you avoid these consequences.

8.13.1 Capitalize proper nouns

Proper nouns include all of the following:

- trademarks
- product names—even if they are not trademarks
- major components or subsystems of a product
  - ▶ The **Holter Telemetry** feature transmits data continuously for a selectable number of hours.
  - ▶ The **Output Delivery System** provides formatting functionality that is not provided by individual procedures.
- other names
  - ▶ International Organization for Standardization
  - ▶ Digital Command Language
- acronyms, initialisms, and short forms of names

8.13.2 Capitalize user-interface labels as they are capitalized in the interface

- When documenting software or any device that has a graphical user interface, use the same capitalization that is used in the interface.
  - ✓ When the **Auto Cap Formation Interval** is set to **Auto**, the formation interval automatically adjusts in order to optimize device longevity and charge times.
- If a term refers to a capitalized user-interface label in some contexts but is used as a common noun in other contexts, then capitalize it only when it refers to the user-interface label:
  - ✓ Select the **Default Theme** that you prefer.
  - ✓ Initially, the OLAP Viewer uses the **default theme** that the Information Delivery Portal uses.

- Unless title-style capitalization is used for all user-interface labels, use a different font or some other typographic convention to visually distinguish the labels from the surrounding text. In the following example, the monospace font is an important syntactic cue that helps readers interpret the sentence correctly:

  ✗ Select the Run to cursor command from the pop-up menu.

  ✓ Select the `Run to cursor` command from the pop-up menu.

8.13.3 Don't capitalize common nouns

In software messages, inappropriate capitalization of common nouns is rampant:

✗ The **Default Aggregation** must contain all **Levels** in a **Cube**.

✓ The **default aggregation** must contain all **levels** in a **cube**.

Sometimes this misguided convention is carried over into documentation:

✗ The **Episode Data** that the device records is summarized below.

✓ The **episode data** that the device records is summarized below.

In addition, many software developers, engineers, and other product developers regard every component or feature that they develop as being so noteworthy that it merits capitalization. As the passage of time renders their creation less special, the capitalization falls by the wayside, leading to inconsistencies.

Here are some guidelines that will help you identify and combat unwarranted capitalization:

- If a name is followed by another noun that is not part of the name, don't capitalize that noun:

  ✓ the Advanced Search **dialog box**

  ✓ the Cardiac Compass **report**

  ✓ the Patient Information **screen**

  However, note that a final noun is often regarded as part of a name:

  ✓ the Model 3650 Edelmann AVR **System**

- If there can be more than one of something, then it probably is not a proper noun and should not be capitalized:

  ✓ application server

  ✓ device status indicator

There are exceptions, but many of the exceptions are clearly geographical or geopolitical names:

✓ the United States

✓ the Great Lakes

✓ the Alps

✓ ActiveX Data Objects

- Conversely, don't assume that something is a proper noun merely because there can be only one of it in a particular context. In the following example, *fact* is not a name. It is merely a type of table.

  X A cube consists of only one **Fact** table, plus multiple dimension tables and other types of tables.

  ✓ A cube consists of only one **fact** table, plus multiple dimension tables and other types of tables.

- The fact that an acronym or initialism exists for a term doesn't mean that the full form of the term should be capitalized:

  ✓ cardiopulmonary resuscitation (CPR)

  ✓ cascading style sheet (CSS)

  ✓ magnetic resonance imaging (MRI)

  However, if the acronym or initialism represents a proper noun, then of course the full form should be capitalized:

  ✓ Common Gateway Interface (CGI) is the name of a particular interface.

  ✓ Lightweight Directory Access Protocol (LDAP) and File Transfer Protocol (FTP) are names of network protocols.

8.13.4 When necessary, use capitalization to improve readability

Some terms should be capitalized to prevent misreading. Inform your localization coordinator that you are capitalizing these terms for that reason.

✓ In order to extract metadata, users must have **Read** access to the metadata records and to the database tables.

✓ Do not allow **Lead** or **Active Can** electrodes to come into contact during a high-voltage therapy.

8.13.5 Establish clear lines of communication with localization coordinators

Recognize that your localization staff often must weigh several factors in order to decide whether to translate a capitalized term or not. Make sure they know who to contact when they have questions about such issues.

Keep track of all decisions that you make regarding capitalization, and be consistent.

Recommended Reading

Meyer, Charles F. 1987. *A Linguistic Study of American Punctuation.* New York: Peter Lang.

Chapter 9

Eliminating Undesirable Terms and Phrases

Introduction to Controlling Terminology

The reasons for excluding certain terms from your documentation are similar to the reasons for following the Global English style guidelines. Table 9.1 lists the main goals and gives examples of terminology restrictions that contribute to achieving those goals:

Table 9.1 Goals for Terminology Restrictions

Goal	Ways to Achieve That Goal
Make documents more comprehensible for non-native speakers	▪ Eliminate unnecessary synonyms and unnecessary variations in phrasing. ▪ Eliminate uncommon non-technical terms that non-native speakers might not understand. ▪ Eliminate idioms and metaphors.

(continued)

Table 9.1 *(continued)*

Make translation more efficient, thereby reducing translation costs	▪ Eliminate unnecessary synonyms and unnecessary variations in phrasing. ▪ Eliminate idioms and metaphors. ▪ Replace terms that have too wide a range of meanings with words that have fewer meanings. ▪ Replace wordy phrases with more-concise alternatives.
Make information more accessible and comprehensible to all readers	▪ Eliminate unnecessary synonyms and unnecessary variations in phrasing.
Improve document quality	▪ Eliminate incorrect or obsolete terms.

Unfortunately, there is no comprehensive list of terms that are unsuitable for Global English. Each organization has its own terminology and must make its own decisions about which terms to deprecate.[1] Therefore, the guidelines in this chapter are merely intended to help you compile your own list of terms.

Tools for Controlling Terminology

In order to follow the guidelines in this chapter, you need some type of software that flags undesirable terms as errors. Except perhaps in the smallest organizations, terminology cannot be managed or controlled effectively without the use of software. Human beings are not well suited for memorizing lists of hundreds or thousands of deprecated terms. Even if your authors were willing and able, requiring them to memorize lists would be a waste of human resources.

If your organization has its own staff of programmers, then they could develop terminology-checking software that flags your deprecated terms as errors. However, even the task of developing specifications for the software and its interfaces would take months. Including any kind of linguistic intelligence in the software (as opposed to merely searching for exact matches of text strings) would require skills that the average programmer doesn't have.

[1] *deprecate*: to designate as incorrect or undesirable.

For many organizations, controlled-authoring software is a simpler, more cost-effective, and more powerful solution. See "Controlled-Authoring Software" in Chapter 1, "Introduction to Global English," for more information.

Where to Store Deprecated Terms

Controlled-authoring software applications typically include some type of repository in which to store information about deprecated terms and approved terms. If you are not yet using controlled-authoring software, you need a repository for the deprecated terms that the guidelines in this chapter will help you identify.

If some of your documentation is translated, then your organization might already have a terminology management system (TMS) that you could use as your repository for deprecated terms. However, such systems have been developed to support translation, not to support controlled authoring. Therefore, significant changes to the TMS data structure, interfaces, and workflow are required.

Initially, many organizations find it easier to store information about deprecated terms in a spreadsheet. Spreadsheet applications such as Microsoft Excel enable you to easily add, delete, and rearrange columns until you have a clear idea of what types of terms and information you want to manage, how you might want to subset your data, and so on. As long as you can export your data as tab-delimited or comma-separated values, you can use that data with controlled-authoring software and with other language technologies.

For more information about managing terminology, see the Bibliography.

Researching Terminology Issues

In order to follow some of the guidelines in this chapter, you need to do some research. Often you can use external resources such as dictionaries and search engines. But sometimes you need to determine how terms are used in your own documentation. Table 9.2 lists some terminology guidelines and the questions arising from these guidelines that you cannot answer without searching your own documentation.

Table 9.2 Guidelines and Questions That Require Searching Your Own Documentation

Guidelines	Questions
9.2 Eliminate obsolete terms	Are there contexts in which some obsolete terms must remain unchanged? Even when products are renamed or replaced, their predecessors often remain available and must be referred to in some contexts.
9.6 Eliminate incorrect technical terms	Are you sure that a term that you are suggesting as an alternative has the exact same meaning and is used in the same contexts? For example, can you be certain that *box chart* is always a suitable replacement for *box graph* without researching those terms?
9.7 Eliminate variant spellings 9.8 Eliminate orthographic variants	When more than one form of a term is used in your documentation, which one should you deprecate?
9.11 Eliminate unnecessary Latin abbreviations 9.12 Eliminate other non-technical abbreviations 9.13 Eliminate clipped terms	Are you sure that none of these terms is used in your documentation to mean something else? For example, does *p.a.* always represent *per annum*, or does it have some other meaning in your documentation? Does *rep* stand for *repetition*, or is it also a clipped form of some other term?
9.15 Eliminate unusual non-technical words 9.16 Eliminate other unnecessary synonyms 9.17 Eliminate wordy phrases 9.19 Eliminate certain idiomatic phrasal verbs	Are you sure that at least one of your proposed alternatives is appropriate in all contexts?

Terminology research requires two things:

- a large, representative collection of your organization's documentation
- a way of retrieving from that collection every occurrence of the terms that you are researching, along with the contexts in which the terms are used

A useful tool for meeting the second requirement is a type of software called a *concordancer*—a search engine that is designed for language study. When you search a collection of documents for a term, the concordancer shows you every instance of that exact term, along with the contexts in which each instance occurs. For information about concordancers, see http://www.globalenglishstyle.com.

Concordancers are relatively easy to develop, so if your organization has programming resources, you might choose to develop your own. If you use controlled-authoring software, that software probably provides a way of extracting specified terms and their contexts from your collection of documents.

The following sections illustrate why research that is based on your own documentation is sometimes required.

Researching Orthographic Variants

As guideline 9.8, "Eliminate orthographic variants," points out, *orthography*[2] is less standardized in English than spelling. Dictionaries don't agree with each other, and they often don't take actual usage into account. There is no single external authority that you can consult, because technical documentation typically contains terms that are not in any dictionary.

In such situations, authors and editors often turn to the Internet. However, their research is hampered by the fact that most search engines disregard case and punctuation. For example, suppose you are curious about how often the word *reenter* is hyphenated on the Internet. If you use Google to search for exact matches of that term, nearly all of the hits

---

[2] *orthography*: how terms are written: as one word or two, with or without a hyphen, capitalized in one way or another.

are valid. But if you search for *re-enter*, Google returns links to texts that contain all of the following variations:

- re enter
- Re-Enter
- Re: enter
- Re: Enter
- ReEnter

Therefore, there is no way to determine how many hits were for *re-enter* rather than for the above variants.

Of course, you should not use the mostly unedited contents of the World Wide Web as the sole basis for decisions about orthography. As mentioned above, you should base such decisions at least in part on how terms have been written in your existing documentation. If *ZIP code* occurs in your documentation 50 times more often than *zip code*, then *ZIP code* is obviously the form that most of your authors consider correct. Similarly, if *multi-row* outnumbers *multirow* by a ratio of 10 to 1, then apparently most of your authors think that the hyphen makes this uncommon prefixed term more readable. Flagging *ZIP code* and *multi-row* as errors would merely create unnecessary work.

However, consistency is the main goal here. You might choose to rely solely on external authorities when possible. But you will still need to search your own documentation in order to make decisions about terms that are not included in any external reference.

Researching Synonyms

Are two terms exact synonyms, and if so, which term should you standardize on? Sometimes you know instinctively (or because you are familiar with your organization's technical terminology) that *term A* can always be replaced by *term B*. For example, you might know that, in your organization, *box chart* and *box graph* are synonyms.

If you aren't sure, you can consult the subject-matter experts (SMEs) in your organization. Be careful about relying solely on SMEs, however. Especially in a large organization, SMEs don't all have the same opinions or the same degree of familiarity with the subject matter. Finding the SME who can give you the most reliable answer can be a challenge.

It is always good to test your own assumptions (and those of your SMEs) by looking at how terms are used in your own documentation. Even if everyone agrees that two terms are exact synonyms, you probably want to know how frequently each term occurs in your documentation. Then you can make a more informed decision about which term to approve and which to deprecate.

9.1 Eliminate trademark violations

Many organizations devote considerable resources to choosing trademarks and to establishing their brand identities, yet they put little effort into ensuring that their trademarks are used correctly and consistently. Here are some common types of trademark violations that can weaken or invalidate a trademark or dilute a brand:

- Making a trademark possessive:
  - ✗ **GraphMagic's new Import wizard**
  - ✓ the new **GraphMagic** Import wizard
- Joining a trademark with a prefix or suffix, or with another word:[3]
  - ✗ **non-Oracle** data types
  - ✓ data types that are not supported by Oracle
  - ✗ a **Java-based** protocol
  - ✓ a protocol that is based on Java
- Using incorrect punctuation or capitalization:
  - ✗ HP/UX
  - ✓ HP-UX (Hewlett-Packard trademark for an operating system)
  - ✗ SYBASE
  - ✓ Sybase
- Using an acronym or initialism instead of spelling out the complete trademark:
  - ✗ The **FM** update is being sent to all registered users.
  - ✓ The **FileMaker** update is being sent to all registered users.
- Using a trademark as a noun or a verb:
  - ✗ Applicants within the U.S. should include a **Xerox** of both sides of their I-94 card.
  - ✓ Applicants within the U.S. should include a **photocopy** of both sides of their I-94 card.
  - ✗ You can **xerox** the Grant Deed form if you need additional copies.
  - ✓ You can **photocopy** the Grant Deed form if you need additional copies.

---

[3] Some organizations tolerate this type of trademark violation, in part because the revisions are often quite awkward.

9.2 Eliminate obsolete terms

Ask the subject-matter experts in your organization to report any changes in terminology that they become aware of. Eliminate all of the following types of terms that occur in your documentation, regardless of whether those terms originate in your own organization:

- obsolete names of companies, organizations, and business units
- obsolete product names
- obsolete Web addresses
- obsolete technical terms

9.3 Eliminate internal terms

Eliminate terms that are appropriate for a technical or internal audience but not for a general audience. For example, many technical communicators are familiar with the term *callout*, but that term is not appropriate for other readers:

✗ In the figure, **callouts** are used to indicate the main parts of the portal.

✓ In the figure, the main parts of the portal are numbered, and explanations are provided below.

Similarly, the *Microsoft Manual of Style for Technical Publications* recommends avoiding several software terms that are familiar to software developers or to systems analysts but not to general users.

Also eliminate temporary, internal names for new products or components. For example, after Microsoft Windows Vista was released, its code name, Longhorn, became obsolete.

9.4 Eliminate text strings that indicate errors in a source file

Your publishing software probably has a small set of text strings that indicate errors in a source file. Some of these text strings don't appear until you generate a particular type of output from the source files.

For example, suppose that you convert XML source files to HTML or PDF output. If the conversion software doesn't find the target of an XML link, the HTML or PDF output typically contains an error message such as *Link target not found.*

If the XML source contains the phrase *on page*, then the XML probably contains a page number instead of a link. Since page numbers are not known until a printable output format is generated, that phrase and the page number are probably an error. Perhaps they were accidentally left in the document when the document was converted from a print-only format.

9.5 Eliminate repeated words and phrases

Grammar-checking software and controlled-authoring software typically detect any two repeated words. You don't need to specify the word sequences explicitly. Here are a few that occur frequently:

- a a
- of of
- the the

Some controlled-authoring applications also detect repeated phrases and ungrammatical sequences of words like the following:

- Themes control the appearance **of the of the** portal Web application.
- The `Results` item is available only after have **you have execute** the Run Event code.
- If you select the `Others` item, the text is parsed as **if is** English.

9.6 Eliminate incorrect technical terms

Inconsistent terminology is confusing. When you use different terms to represent the same technical concept, readers often cannot find the information that they need.

Incorrect or Non-standard	Correct or Standard[4]
box graph	box chart
message area	status bar
shortcut menu	pop-up menu
preset style	predefined style

9.7 Eliminate variant spellings

Variant spellings make translation less efficient because they contribute to unnecessary variation in your documentation.

9.7.1 Non-standard, informal spellings

Non-standard	Standard
hi	high
lite	light
lo	low
thru	through

9.7.2 Spellings from other varieties of English

Some English terms are spelled differently in different parts of the world. Use the spellings that are most appropriate for the majority of your readers.

[4] Not everyone agrees that these terms are correct or that the terms in the first column are incorrect. They are listed here as possible examples.

Here are some terms that are spelled differently in the United Kingdom (and probably in some other countries as well) than in the U.S.:

British spelling	American spelling
aluminium	aluminum
cancelled, cancelling	canceled, canceling
catalogue	catalog
colour	color
grey	gray

 **Apparent Misspellings Might Actually Be Correct**

Sometimes different spellings of the same term are considered correct in different subject areas. For example, the *Wall Street Journal* refers to U.S. Treasury bonds as *treasurys* rather than as *treasuries*.

9.8 Eliminate orthographic variants

Technically, differences in capitalization and hyphenation and in whether a term is written as one word or two are matters of orthography, not of spelling. In English, orthography is less standardized and more likely to change than spelling.

9.8.1 Terms that can be written either as one word or two

Many compound nouns are written as two words when they are first introduced and as one word when they become more familiar. The one-word form might become common in some subject areas before it is adopted in other subject areas.

For example, *Merriam-Webster's Online Dictionary* has an entry for the noun *time stamp* (a mechanical device that stamps the date and time on letters and papers). But if you search Google, *timestamp* and *timestamps* outnumber *time stamp* and *time stamps* by a ratio of nearly 8 to 1. The one-word form is especially prevalent in the context of computer software.

When you must decide whether to write a term as one word or two, consider following this approach:

- Search online dictionaries and glossaries—especially any that pertain specifically to your subject matter. The OneLook Dictionary Search Site (http://www.onelook.com) enables you to consult a large number of dictionaries at once. Printed dictionaries and glossaries are fine, too, of course, but they aren't as easy to search.
- Search the Web to determine which variant is most common.
- Search your own collection of documentation, if possible. If one variant of a term greatly outnumbers another in your legacy documentation, is it worth the effort to standardize on the less-common variant? It is most important to standardize nouns and verbs that your readers are likely to search for.

Here are some other terms for which both one- and two-word forms are common:

- healthcare, health care
- roadmap, road map
- wildcard, wild card

9.8.2 Terms that are capitalized incorrectly

As mentioned in Chapter 8, "Punctuation and Capitalization," inappropriate or inconsistent capitalization interferes with the use of language technologies such as translation memory, machine-translation software, and controlled-authoring software. In addition, human translators need to know whether a noun is indeed a name (proper noun) or just a common noun. In some languages and contexts, a term is translated differently depending on whether it is a proper noun or a common noun. See section 8.13 for capitalization guidelines.

9.8.3 Terms that use non-standard hyphenation

Inconsistent hyphenation interferes with the use of language technologies. See section 8.7 for hyphenation guidelines.

9.8.4 Non-standard transliterations

There are standard transliterations for many terms such as *algebra*, *glasnost*, and *sushi* that entered the English language from languages that use different writing systems. For other terms (including names of individuals, places, and companies), many different transliterations are in use. For example, the last name of the Libyan leader Moammar Gadafy has also been transliterated as Gaddafi, Gadafi, Qadahi, Khadafi, Khadafy, and probably at least half a dozen other ways.

If you use such terms in your documentation, then decide which transliterations you will use.

9.8.5 Terms that include foreign characters or diacritics

To avoid unnecessary variation, decide whether you will use the appropriate characters or diacritics in terms that have been borrowed from other languages that use the Latin alphabet.

Without Diacritics	With Diacritics
facade	façade
Mobius strip	Möbius strip
moire pattern	moiré pattern

9.9 Eliminate terms from other varieties of English

Some organizations produce different versions of their English documentation for different English-speaking markets, but most organizations standardize on one variety of English. In either case, you need to be aware of differences in terminology in addition to differences in spelling (guideline 9.7.2).

If you search the Internet for all or some of the words *differences English American British Canadian Indian Australian* (for example), you will find that most of the differences are in vocabulary that doesn't occur very often in most technical documentation. For example, one source states that Canadians *turn on the* ***tap***, whereas Americans *turn on the* ***faucet***.[5] The British phrase *make someone redundant* is approximately equivalent to the American phrase *lay someone off*.

[5] Cornerstone Word Company, http://www.cornerstoneword.com/misc/cdneng/cdneng.htm.

On the other hand, Indian English includes some *-ion* nouns that don't exist in U.S. English and that could conceivably occur in technical documents. Unless your main audience is in India, you should eliminate such terms:

✗ The getDateModified class returns the date and time of **updation** of the metadata.

✓ The getDateModified class returns the date and time on which the metadata **was updated**.

✗ The textile industry needs to invest heavily in the modernization and **upgradation** of existing manufacturing facilities.

✓ The textile industry needs to invest heavily in the modernization and **upgrading** of existing manufacturing facilities.

9.10 Eliminate obscure foreign words

There is no need to ban common foreign words such as *per*, *via*, and *vice versa*[6] for which there are no good alternatives. However, you can easily replace some foreign terms with common Anglo-Saxon words:

✗ The getDelegateClassName method returns the name of the delegate class, **sans** package name.

✓ The getDelegateClassName method returns the name of the delegate class, **without the** package name.

In other cases, you can simply eliminate the foreign term or use a more substantial revision:

✗ Graphs are not required **per se**, but you must include either a table object or a graph object in your report.

✓ Graphs are not required. However, you must include either a table object or a graph object in your report.

[6] *Via* can be replaced by *by way of* in some contexts, but in many contexts that revision seems unnatural. *En route* is another Latin phrase that you might choose to allow, at least in some contexts. *Sic* will baffle many non-native speakers, but there is no good alternative.

9.11 Eliminate unnecessary Latin abbreviations

Because they are not used by all authors in all contexts, Latin abbreviations are a significant source of inconsistency. In addition, not all non-native speakers are familiar with them.

As in guideline 9.10, you don't need to banish common Latin abbreviations such as *a.m.* or *p.m.*, or abbreviations such as *ibid.*, *op cit.*, and *et al.* that are standard in some scientific and technical publications. However, use the Anglo-Saxon equivalents of the following abbreviations whenever possible:

Latin abbreviation	Anglo-Saxon equivalent
ca.	about, approximately
e.g.	for example
etc.	and so on
i.a.	among others
i.e.	that is
p.a.	per year, yearly

9.12 Eliminate other non-technical abbreviations

Don't assume that a non-native speaker will understand a non-technical abbreviation whose meaning is obvious to you. Some abbreviations are puzzling to translators as well.

Abbreviation	Full form
a.k.a., AKA	also known as
Assn.	Association
eval.	evaluation evaluate
n.a., N.A., n/a, N/A	not applicable not available none
oz.	ounces[7]

[7] Most of the world uses the metric system. If you use English units of measurement in your documentation, then include metric equivalents as well.

Note: Include abbreviations of *technical* terms in your published glossary unless they are standard abbreviations that your main audience understands. Include *all* abbreviations in your translation glossary. See guideline 3.7.1, "Consider defining or explaining noun phrases," for a discussion of translation glossaries.

9.13 Eliminate clipped terms

In a clipped term, the full form of the term is truncated, but no period is used. Like abbreviations, clipped terms can be puzzling to non-native speakers and translators. They cause unnecessary variation, and many of them are too informal for most technical documents.

Short form	Full form
app	application
autocreate	automatically create
dupe	duplicate
flu	influenza
org chart	organizational chart
quote (noun)	quotation mark
specs	specifications
stats	statistics
vet	veterinarian

Suggesting Alternatives for Abbreviations and Clipped Terms

When you compile information about abbreviations and clipped terms that you want to deprecate, make sure that your suggested alternatives include all of the full forms that each of these short forms might represent in your documentation. For example, *rep* could be the clipped form of *repetition*, *representative*, and perhaps even *repeat, represent*, and *repertory*. That, of course, is another good reason for deprecating such terms.

9.14 Eliminate certain contractions

Anyone who is reasonably proficient in English certainly recognizes and understands common contractions such as *isn't* and *won't*. However, don't use unusual contractions that non-native speakers are unlikely to understand. Also eliminate contractions that are too informal for the types of technical documentation that you produce.

Here are some specific suggestions:

- Don't use the following contractions. They are unusual, non-standard, or excessively informal:

  X ain't, could've, mightn't, might've, mustn't, must've, shan't, should've, that'll, 'tis, 'twas, would've

- Don't use contractions of interrogatives:

  X how'd, how'll, how's, what'd, what's, when'd, when'll, when's, where'd, where'll, where's, who'd, who'll, who's, why'd, why'll, why's

- Decide whether the following contractions are suitable for your documentation:
  - Contractions of first person, third-person masculine, and third-person feminine pronouns:[8]

    ? he'd, he'll, he's, I'd, I'll, I'm, I've, she'd, she'll, she's, we'd, we'll, we're, we've

  - Common contractions that might be too informal for many types of technical documentation:

    ? aren't, can't, couldn't, didn't, doesn't, don't, hadn't, hasn't, haven't, isn't, it's, shouldn't, they'd, they'll, they're, they've, wasn't, weren't, won't, wouldn't, you'd, you'll, you're, you've

    ? here's, let's, that's, there's

As always, consistency is important. If some authors in your organization use *can't*, others use *can not*, and others use *cannot*, then you have a lot of unnecessary variation in your documentation.

---

[8] The issue here is not only whether the contractions are appropriate, but also whether the pronouns are appropriate in your documentation.

Dealing with Ambiguous Contractions

The *'d* and *'s* contractions each have two meanings: *'d* can represent either *had* or *would*, and *'s* can represent either *is* or *has*. To human readers, the intended meaning is usually clear from the context. If you are using machine-translation software, or if you are concerned about this ambiguity for other reasons, then use *'d* only to represent *would*, and use *'s* only to represent *is*. Those meanings are the most common in technical documentation.

If you are using controlled-authoring software, then you can use the linguistic context to determine how *'d* and *'s* are being used. If *'d* is followed by the base form of a verb, then it represents *would*:

✓ **They'd** never believe me.

If *'d* is followed by *had*, *been*, or some other past participle, then it represents *had*:

▸ **They'd** never had so much fun before.

▸ **They'd** never been so happy.

▸ **They'd** never thought much about retirement.

Similarly, if *'s* is followed by an adjective, a noun phrase, or the -ING or past participle form of a verb, then it represents *is*:

✓ **It's** not easy being pretty.

✓ **He's** a great guy.

✓ **It's** going to be a long, hot summer.

✓ **It's** believed to be the first time a nine-year-old has ever won this competition.

If *'s* is followed by *had* or by *been*, then it represents *has*:

▸ **He's** had a dozen job offers since he was laid off.

▸ **It's** been a long, hot summer.

9.15 Eliminate unusual non-technical words

Some unusual non-technical words, such as *albeit*, are unfamiliar to many non-native speakers. Replace such terms with more-common alternatives.

Unusual	Common
albeit	although, but
amongst	among
hereinafter	in the rest of this document
inordinately	extremely, unusually, too
whilst (mainly British)	while

9.16 Eliminate other unnecessary synonyms

Guideline 9.15 discussed unusual non-technical words, but even some relatively *common* words and phrases have synonyms that are *more* common. You can reduce unnecessary variation by standardizing on the more-common equivalents.

From a native speaker's perspective, there is nothing wrong with the less-common words and phrases in the table below. However, remember that every unnecessary inconsistency adds to the cost of translation. In addition, some non-native speakers are not familiar with some of the less-common synonyms.

Less common	More common
and so forth	and so on
Apart from that	In addition
as though	as if
for instance	for example
Though	Although
, though	, although , but

If two synonyms occur with about the same frequency, then standardize on the one that has the smaller range of meanings. For example, with the verbs *determine*, *know*, and *see*, you might choose to use *whether* instead of *if*:

✗ To **determine if** a terminal supports automatic left margins, perform the following test:

✓ To **determine whether** a terminal supports automatic left margins, perform the following test:

Similarly, *information about* seems more literal and precise than *information on*, because *about* has a smaller range of meanings in technical documentation:

✗ For more **information on** data modeling, see "An Overview of Data Modeling" on page 46.

✓ For more **information about** data modeling, see "An Overview of Data Modeling" on page 46.

9.17 Eliminate wordy phrases

To reduce word counts and simplify your syntax, replace wordy phrases with single words or with shorter phrases.

Wordy	Not Wordy
a great many	many
a number of	many, several
at the present time	now, yet
click the OK button	click OK
despite the fact that	although, even though
from time to time	periodically, occasionally
have the ability to	be able to, can
in between	between
in many cases	often
once in a while	occasionally, sometimes
quite a few	several
whether or not	whether

In particular, don't use a verb + noun if you could use just a verb instead:

Wordy Verb + Noun	Not Wordy
come to a conclusion	conclude
conduct an investigation	investigate
make a decision	decide
reach an agreement	agree

See guideline 3.3, "Use a verb-centered writing style," for additional examples and discussion.

9.18 Eliminate idioms

An idiom is a group of words whose meaning is not derived from the literal meanings of the individual words. Non-native speakers and translators cannot be expected to understand all of the idioms that are used in English.

Technical documents rarely contain blatantly idiomatic phrases like the following:

- in a nutshell
- the whole nine yards
- the bottom line
- the eleventh hour

However, *keep an eye out* for more-subtle idioms:

✗ Use a four-pin Y cable splitter if you **are short on** four-pin connectors from your power supply.

✓ Use a four-pin Y cable splitter if you **don't have enough** four-pin connectors from your power supply.

✗ **For the most part**, corresponding windows in other operating environments show similar results.

✓ Corresponding windows in other operating environments **generally** show similar results.

✗ When displaying the precedence relationships between activities on the Gantt chart, **bear in mind** the following facts:

✓ When displaying the precedence relationships between activities on the Gantt chart, **consider** the following facts:

9.19 Eliminate certain idiomatic phrasal verbs

Some idiomatic phrasal verbs occur more frequently in technical documentation than other idioms. If a phrasal verb is very common, then many non-native speakers and most professional translators will understand it. For example, many non-native speakers will understand *figure out*:

▶ The first step is to **figure out** how to extract the data.

On the other hand, you can easily use the non-idiomatic, one-word verb *determine* instead of the idiomatic *figure out*:

✓ The first step is to **determine** how to extract the data.

Therefore, *figure out* is an unnecessary synonym for *determine*, and it contributes to unnecessary variation in your documentation.

In the next example, *represent* is a good substitute for the idiomatic verb phrase *stand for*:

✗ Negative numbers **stand for** the bottom *n* values.

✓ Negative numbers **represent** the bottom *n* values.

As always, consider whether more-extensive revisions would be suitable—especially if you can reduce word counts at the same time.

✗ If you **come across** a document with tables that are wider than the screen, press PageDown. (16 words)

✓ If a document's tables are wider than the screen, press PageDown. (11 words)

✗ The details of how user exits are implemented are **left up to** the operating system to determine. (18 words)

✓ The details of how user exits are implemented are **determined by** the operating system. (14 words)

✓+ User exits are implemented differently on different operating systems. (9 words)

 **Phrasal Verbs That Don't Have Single-Word Equivalents**

Don't try to ban all idiomatic phrasal verbs, because some of them don't have non-idiomatic, single-word equivalents. In the following example, there is no one-word verb that you could use instead of *catch up*:

▶ Severe weather might reduce production during one period. However, in the following period production might increase as producers struggle to **catch up**.

Other idiomatic phrasal verbs don't have single-word alternatives that are suitable in all contexts. For example, sometimes *consider* is an acceptable substitute for *keep in mind*:

▶ Four Things to **Keep in Mind** When Choosing a Hybrid Car

✓ Four Things to **Consider** When Choosing a Hybrid Car

But *consider* doesn't have quite the same meaning as *keep in mind*, and it is not a good substitute in some contexts:

▶ Please **keep in mind** that this release is still under development.

⦸ Please **consider** that this release is still under development.

9.20 Eliminate colloquialisms

Colloquialisms are words or phrases that are acceptable in casual speech but not in formal speech or writing. They contribute to unnecessary variation and are not appropriate for global communication. Here are some examples:

✗ A multicast is **kind of like** the packet types that were discussed in Chapter 2.

✓ A multicast is **similar to** the packet types that were discussed in Chapter 2.

✗ See the Support Web site for **lots more** annotated sample programs.

✓ See the Support Web site for **additional** annotated sample programs.

✗ The 1.2-inch external display is **a hair smaller than** the VX8100 display, but it has the same resolution (65,000 colors).

✓ The 1.2-inch external display is **slightly smaller than** the VX8100 display, but it has the same resolution (65,000 colors).

✗ **What's more**, two of the early Web 1.0 exemplars were pioneers in treating the Web as a platform.

✓ **Moreover**, two of the early Web 1.0 exemplars were pioneers in treating the Web as a platform.

9.21 Eliminate metaphors

In a metaphor, a term or expression is used in a non-literal sense in order to suggest a similarity. Many non-native speakers understand common metaphors. For example, the following metaphors are common in speech and are seldom thought of as being metaphorical:

▶ **A handful of** companies have refined VCSEL technology so that it is cheaper, faster, and more versatile.

▶ The topic was **tabled** until more input could be obtained from Marketing.

Nevertheless, translation is easier if you *stick to* literal uses of language:

✓ **A few** companies have refined VCSEL technology so that it is cheaper, faster, and more versatile.

✓ The topic was **postponed** until more input could be obtained from Marketing.

Uncommon metaphors add zest to your writing, but they baffle non-native speakers, and they are very difficult to translate:

✗ The new model was designed to overcome the limitations of the **vanilla** CALS model.

✓ The new model was designed to overcome the limitations of the **standard** CALS model.

✗ You must install the Java plug-in on client machines before running applets that require it. This requirement is **palatable** for installed products or intranet solutions, but **the price may be too high** for some thin-client reports.

✓ If you use a Java plug-in, then you must install the plug-in on client machines before running applets that require it. This requirement is **not a problem** for installed products or intranet solutions, but **you might want to use a different approach** for some thin-client reports.

✗ A closer look **under the hood** reveals that each browser supports different implementations of the Java Runtime Environment.

✓ Each browser supports different implementations of the Java Runtime Environment.

Related Guidelines

Chapter 2, "Conforming to Standard English," contains several guidelines that are actually terminology guidelines but that were also appropriate for that chapter:

- 2.2, "Use nouns as nouns, verbs as verbs, and so on"
- 2.3, "Don't add verb suffixes or prefixes to nouns, acronyms, initialisms, or conjunctions"
- 2.4, "Use standard verb complements"
- 2.5, "Don't use transitive verbs intransitively, or vice versa"
- 2.6, "Use conventional word combinations and phrases"
- 2.7, "Don't use non-standard comparative and superlative adjectives"

Appendix A

Examples of Content Reduction

Introduction

As noted in Chapter 1, "Introduction to Global English," content reduction is one of the best ways to reduce translation costs. If you understand both your subject matter and your audience, you might be able to identify entire topics or sections that are superfluous.

As the examples here show, even if you focus only on sentences and paragraphs, you can often decrease your word counts by 20% or more. If your document will be translated into several languages, at an average cost of about $.25 per word for each language, your cost savings can be significant.

Note: In the "before" versions of these examples, a gray background indicates text that was deleted or revised. Footnotes are in the Commentary sections rather than at the bottom of the page. The words in the topic titles are not included in the word counts.

Example 1

Before	After	Reduction
22 words	15 words	**32%**

Before:

To create a link, type the link text in the `Text` field and click `Add`. The message "Link was successfully added" appears.

After:

To create a link, type the link text in the `Text` field and click `Add`.

Commentary

As Rushanan (2007) points out, it is not necessary to document everything that happens in a user interface. In addition, it is best not to quote system messages. Suppose that translator A translated the software (including the "Link was successfully added" message), and that translator B is now translating the documentation. Translator B has to locate translator A's translation in order to ensure that she translates "Link was successfully added" in exactly the same way. Moreover, if a software developer changes the text of the message, the translation must be changed both in the software and in the documentation.

Such details are necessary only if the main purpose of the document is to facilitate software testing. In that case, testers need to know all the details about what users are supposed to see in order to verify that everything works correctly.

Example 2

Before	After	Reduction
63 words	48 words	**24%**

Before:

Specify Conditional Highlighting for Graph Values

To specify conditional highlighting for values in a graph,[1] complete these steps either in the `Layout` section of the Edit Report view or in the View Report view:

1. On the graph toolbar, click `Specify Conditional Highlighting for Graph Values`.
2. From the pop-up menu that appears,[2] select `Conditional Highlighting`.
3. Complete the Conditional Highlighting dialog box.
4. When you are finished,[3] click `OK`.
5. Save the report.

After:

Specify Conditional Highlighting for Graph Values

Complete these steps either in the `Layout` section of the Edit Report view or in the View Report view:

1. On the graph toolbar, click `Specify Conditional Highlighting for Graph Values`.
2. From the pop-up menu, select `Conditional Highlighting`.
3. Complete the Conditional Highlighting dialog box.
4. Click `OK`.
5. Save the report.

Commentary

1. The introductory phrase repeats the information that is in the topic title.
2. "That appears" is unnecessary.
3. Users don't need to be told to finish what they are doing before clicking `OK`.

Example 3

Before	After	Reduction
82 words	48 words	**41%**

Before:

Combine Filters

Note: In order to combine filters, you must have assigned filters to at least two data items.[1]

Note: The default filter combination is `Match all filters`.[2]

To combine filters, complete these steps:[3]

1. In the `Data` section of the Edit Report view, click `Advanced`.
2. In the Advanced Query Options dialog box,[4] complete the `Filter combination` section.
3. When you are finished,[5] click `OK`.
4. Save the report.

Note: Only authorized users can save reports. If you have questions about your authorization, contact your system administrator.

After:

Combine Filters

1. In the `Data` section of the Edit Report view, click `Advanced`.
2. Complete the `Filter combination` section.

Note: The default filter combination is `Match all filters`.

3. Click `OK`.
4. Save the report.

Note: Only authorized users can save reports. If you have questions about your authorization, contact your system administrator.

Commentary

1. This note is unnecessary. The audience understands that in order to *combine* filters, they have to have at least two filters.

2. If the `Match all filters` default value appears in the interface, then this note is also unnecessary.

3. The introductory phrase repeats information that is in the topic title.

4. Specifying the title of every window or dialog box that appears is expensive when your content will be translated into multiple languages.

5. Users don't need to be told to finish what they are doing before clicking `OK`.

Example 4

Before	After	Reduction
98 words	72 words	**27%**

Before:

A manual adjustment can be applied either before the rule-generated adjustments or after the rule-generated adjustments, in accordance with your specifications. The rule-generated adjustments run in an order that you specify. The complete adjustment process has these three phases:

1. Apply all the "Before Rules" manual adjustments. These adjustments are listed on the `Before Rules` tab of the Adjustments window.

2. Run the adjustment rules in their listed order and apply all the resulting rule-generated adjustments.

3. Apply all the "After Rules" manual adjustments. These adjustments are listed on the `After Rules` tab of the Adjustments window.

After:

To apply one or more manual adjustments, use the Adjustments window. Specify adjustments in the following order:

1. On the `Before Rules` tab, specify all the manual adjustments that you want to apply before the rule-generated adjustments.
2. Run the adjustment rules in their listed order, and apply the resulting rule-generated adjustments.
3. On the `After Rules` tab, specify all the manual adjustments that you want to apply after the rule-generated adjustments.

Commentary

The "before" version was reorganized substantially. The "after" version is much easier to read and comprehend.

Appendix B

Prioritizing the Global English Guidelines

Introduction

The Global English style guidelines in Chapters 2–7 could have been grouped and ordered in any number of ways. As you can tell from the chapter titles, they are grouped thematically. However, for each guideline, a Priority line indicates how important or useful the guideline is for different audiences or purposes. Here is an example:

Priority: HT1, NN2, MT3

The following tables explain the acronyms and the priority levels:

Acronym	Meaning
HT	human translation
NN	non-native speakers
MT	machine translation

Priority Level	Meaning
1	high priority
2	medium priority
3	low priority

In other words, HT1 indicates that the guideline has high priority for documents that will be translated by human translators. NN2 indicates that the guideline has medium priority for untranslated documents that will be read in English by non-native speakers. And MT3 indicates that the guideline has low priority for documents that will be translated using machine-translation software. These priority values are based on the author's subjective assessments and on feedback from translators and other localization professionals.

In this appendix, the priority values have been used to sort the guidelines and to thereby help you decide which guidelines to focus on—in case you aren't prepared to adopt or implement all of them!

- If your documents will be translated, then you will probably be most interested in the tables that are sorted by the values of HT. See "Prioritized for Translation by Human Translators" on page 239.
- If your documents will not be translated, but you want to make them more suitable for a global audience that includes non-native speakers, then you will probably be most interested in the tables that are sorted by the values of NN. See "Prioritized for Non-Native Speakers of English" on page 242.
- If you use machine-translation software, then you need more information than this book provides.[1] However, the tables that are sorted by the values of MT are at least a starting point for discussions with your software vendor. See "Prioritized for Machine Translation" on page 246.

Note: Priority values were not assigned to the punctuation and capitalization guidelines in Chapter 8 or to the terminology guidelines in Chapter 9. Therefore, those guidelines are not included in the tables.

[1] As explained in "Machine-Translation Software" in Chapter 1, these guidelines are necessary but not sufficient for getting high-quality translations from machine-translation software.

Prioritized for Translation by Human Translators

If your documentation is translated by human translators rather than by machine-translation software, then the guidelines that have a priority of 1 in the HT column are the most important.

Guidelines that are helpful to non-native speakers are also important, because you surely don't translate all of your documents into all of the world's languages. For example, even within the United States, you might have readers whose first language is Vietnamese, Arabic, or Turkish. Most organizations don't translate all of their documentation into all of those languages. Therefore, in the following tables, the guidelines are sorted first by HT (human translation), and then by NN (non-native speakers).

Sorted in a Single Table

Number	Description	HT	NN
2.1	Be logical, literal, and precise in your use of language	1	1
3.1	Limit the length of sentences	1	1
6.1	Don't use a telegraphic writing style	1	1
3.7	Consider defining, explaining, or revising noun phrases	1	2
6.11	Make each sentence syntactically and semantically complete	1	2
2.3	Don't add verb suffixes or prefixes to nouns, acronyms, initialisms, or conjunctions	1	3
7.3	Revise dangling -ING phrases	1	3
2.2	Use nouns as nouns, verbs as verbs, and so on	2	2
2.4	Use standard verb complements	2	2
3.11	Avoid ambiguous verb constructions	2	2
3.12	Write positively	2	2
3.3	Use a verb-centered writing style	2	2
4.2	Clarify what each prepositional phrase is modifying	2	2
4.3	Clarify what each relative clause is modifying	2	2
5.1	Make sure readers can identify what each pronoun is referring to	2	2
6.3	Use *that* with verbs that take noun clauses as complements	2	2
6.4	Use *that* in relative clauses	2	2
6.5	Clarify which parts of a sentence are being joined by *and* or *or*	2	2
6.6	Revise past participles	2	2
6.7	Revise adjectives that follow nouns	2	2

(*continued*)

7.1	Revise -ING words that follow and modify nouns	2	2
7.4	Punctuate -ING phrases correctly	2	2
7.5	Hyphenate -ING words in compound modifiers	2	2
7.6	Eliminate unnecessary -ING phrases and -ING clauses	2	2
3.6	Limit your use of passive voice	2	3
3.8	Use complete sentences to introduce lists	2	3
4.4	Use *that* in restrictive relative clauses	2	3
4.6	Clarify ambiguous modification in conjoined noun phrases	2	3
5.2	Don't use *this*, *that*, *these*, and *those* as pronouns	2	3
5.3	Don't use *which* to refer to an entire clause	2	3
6.2	In a series of noun phrases, consider including an article in each noun phrase	2	3
3.10	Avoid unusual constructions	3	2
6.8	Use *to* with indirect objects	3	2
7.2	Revise -ING words that follow certain verbs	3	2
2.5	Don't use transitive verbs intransitively, or vice versa	3	3
2.6	Use conventional word combinations and phrases	3	3
2.7	Don't use non-standard comparative and superlative adjectives	3	3
2.8	Use *the* only with definite nouns	3	3
2.9	Use singular and plural nouns correctly	3	3
3.2	Consider dividing shorter sentences	3	3
3.4	Keep phrasal verbs together	3	3
3.5	Use short, simple verb phrases	3	3
3.9	Avoid interrupting sentences	3	3
4.1	Place *only* and *not* immediately before whatever they are modifying	3	3
4.5	Consider moving anything that modifies a verb to the beginning of the clause or sentence	3	3
6.10	Consider using *if. . . then*	3	3
6.9	Consider using *both . . . and* and *either . . . or*	3	3
7.7	Revise ambiguous -ING + noun constructions	3	3
7.8	Revise ambiguous *to be* + -ING constructions	3	3

Grouped by Chapter

Number	Description	HT	NN
2.1	Be logical, literal, and precise in your use of language	1	1
2.3	Don't add verb suffixes or prefixes to nouns, acronyms, initialisms, or conjunctions	1	3
2.2	Use nouns as nouns, verbs as verbs, and so on	2	2
2.4	Use standard verb complements	2	2
2.5	Don't use transitive verbs intransitively, or vice versa	3	3
2.6	Use conventional word combinations and phrases	3	3
2.7	Don't use non-standard comparative and superlative adjectives	3	3
2.8	Use *the* only with definite nouns	3	3
2.9	Use singular and plural nouns correctly	3	3

3.1	Limit the length of sentences	1	1
3.7	Consider defining, explaining, or revising noun phrases	1	2
3.3	Use a verb-centered writing style	2	2
3.11	Avoid ambiguous verb constructions	2	2
3.12	Write positively	2	2
3.6	Limit your use of passive voice	2	3
3.8	Use complete sentences to introduce lists	2	3
3.10	Avoid unusual constructions	3	2
3.2	Consider dividing shorter sentences	3	3
3.4	Keep phrasal verbs together	3	3
3.5	Use short, simple verb phrases	3	3
3.9	Avoid interrupting sentences	3	3

4.2	Clarify what each prepositional phrase is modifying	2	2
4.3	Clarify what each relative clause is modifying	2	2
4.4	Use *that* in restrictive relative clauses	2	3
4.6	Clarify ambiguous modification in conjoined noun phrases	2	3
4.1	Place *only* and *not* immediately before whatever they are modifying	3	3
4.5	Consider moving anything that modifies a verb to the beginning of the clause or sentence	3	3

(continued)

5.1	Make sure readers can identify what each pronoun is referring to	2	2
5.2	Don't use *this*, *that*, *these*, and *those* as pronouns	2	3
5.3	Don't use *which* to refer to an entire clause	2	3

6.1	Don't use a telegraphic writing style	1	1
6.11	Make each sentence syntactically and semantically complete	1	2
6.3	Use *that* with verbs that take noun clauses as complements	2	2
6.4	Use *that* in relative clauses	2	2
6.5	Clarify which parts of a sentence are being joined by *and* or *or*	2	2
6.6	Revise past participles	2	2
6.7	Revise adjectives that follow nouns	2	2
6.2	In a series of noun phrases, consider including an article in each noun phrase	2	3
6.8	Use *to* with indirect objects	3	2
6.9	Consider using *both . . . and* and *either . . . or*	3	3
6.10	Consider using *if . . . then*	3	3

7.3	Revise dangling -ING phrases	1	3
7.1	Revise -ING words that follow and modify nouns	2	2
7.4	Punctuate -ING phrases correctly	2	2
7.5	Hyphenate -ING words in compound modifiers	2	2
7.6	Eliminate unnecessary -ING phrases and -ING clauses	2	2
7.2	Revise -ING words that follow certain verbs	3	2
7.7	Revise ambiguous -ING + noun constructions	3	3
7.8	Revise ambiguous *to be* + -ING constructions	3	3

Prioritized for Non-Native Speakers of English

If your documentation is not translated and is read by non-native speakers of English, then the most important Global English guidelines are those that have a priority of 1 in the NN column. Guidelines that are helpful to human translators are less important.

Therefore, in the following tables, the guidelines are sorted first by NN (non-native speakers), and then by HT (human translation).

Sorted in a Single Table

Number	Description	NN	HT
2.1	Be logical, literal, and precise in your use of language	1	1
3.1	Limit the length of sentences	1	1
6.1	Don't use a telegraphic writing style	1	1
3.7	Consider defining, explaining, or revising noun phrases	2	1
6.11	Make each sentence syntactically and semantically complete	2	1
2.2	Use nouns as nouns, verbs as verbs, and so on	2	2
2.4	Use standard verb complements	2	2
3.3	Use a verb-centered writing style	2	2
3.11	Avoid ambiguous verb constructions	2	2
3.12	Write positively	2	2
4.2	Clarify what each prepositional phrase is modifying	2	2
4.3	Clarify what each relative clause is modifying	2	2
5.1	Make sure readers can identify what each pronoun is referring to	2	2
6.3	Use *that* with verbs that take noun clauses as complements	2	2
6.4	Use *that* in relative clauses	2	2
6.5	Clarify which parts of a sentence are being joined by *and* or *or*	2	2
6.6	Revise past participles	2	2
6.7	Revise adjectives that follow nouns	2	2
7.1	Revise -ING words that follow and modify nouns	2	2
7.4	Punctuate -ING phrases correctly	2	2
7.5	Hyphenate -ING words in compound modifiers	2	2
7.6	Eliminate unnecessary -ING phrases and -ING clauses	2	2
3.10	Avoid unusual constructions	2	3
6.8	Use *to* with indirect objects	2	3
7.2	Revise -ING words that follow certain verbs	2	3
2.3	Don't add verb suffixes or prefixes to nouns, acronyms, initialisms, or conjunctions	3	1
7.3	Revise dangling -ING phrases	3	1
3.6	Limit your use of passive voice	3	2
3.8	Use complete sentences to introduce lists	3	2

(*continued*)

4.4	Use *that* in restrictive relative clauses	3	2
4.6	Clarify ambiguous modification in conjoined noun phrases	3	2
5.2	Don't use *this*, *that*, *these*, and *those* as pronouns	3	2
5.3	Don't use *which* to refer to an entire clause	3	2
6.2	In a series of noun phrases, consider including an article in each noun phrase	3	2
2.5	Don't use transitive verbs intransitively, or vice versa	3	3
2.6	Use conventional word combinations and phrases	3	3
2.7	Don't use non-standard comparative and superlative adjectives	3	3
2.8	Use *the* only with definite nouns	3	3
2.9	Use singular and plural nouns correctly	3	3
3.2	Consider dividing shorter sentences	3	3
3.4	Keep phrasal verbs together	3	3
3.5	Use short, simple verb phrases	3	3
3.9	Avoid interrupting sentences	3	3
4.1	Place *only* and *not* immediately before whatever they are modifying	3	3
4.5	Consider moving anything that modifies a verb to the beginning of the clause or sentence	3	3
6.9	Consider using *both . . . and* and *either . . . or*	3	3
6.10	Consider using *if . . . then*	3	3
7.7	Revise ambiguous -ING + noun constructions	3	3
7.8	Revise ambiguous *to be* + -ING constructions	3	3

Grouped by Chapter

Number	Description	NN	HT
2.1	Be logical, literal, and precise in your use of language	1	1
2.2	Use nouns as nouns, verbs as verbs, and so on	2	2
2.4	Use standard verb complements	2	2
2.3	Don't add verb suffixes or prefixes to nouns, acronyms, initialisms, or conjunctions	3	1
2.5	Don't use transitive verbs intransitively, or vice versa	3	3
2.6	Use conventional word combinations and phrases	3	3
2.7	Don't use non-standard comparative and superlative adjectives	3	3
2.8	Use *the* only with definite nouns	3	3
2.9	Use singular and plural nouns correctly	3	3

3.1	Limit the length of sentences	1	1
3.7	Consider defining, explaining, or revising noun phrases	2	1
3.3	Use a verb-centered writing style	2	2
3.11	Avoid ambiguous verb constructions	2	2
3.12	Write positively	2	2
3.10	Avoid unusual constructions	2	3
3.6	Limit your use of passive voice	3	2
3.8	Use complete sentences to introduce lists	3	2
3.2	Consider dividing shorter sentences	3	3
3.4	Keep phrasal verbs together	3	3
3.5	Use short, simple verb phrases	3	3
3.9	Avoid interrupting sentences	3	3

4.2	Clarify what each prepositional phrase is modifying	2	2
4.3	Clarify what each relative clause is modifying	2	2
4.4	Use *that* in restrictive relative clauses	3	2
4.6	Clarify ambiguous modification in conjoined noun phrases	3	2
4.1	Place *only* and *not* immediately before whatever they are modifying	3	3
4.5	Consider moving anything that modifies a verb to the beginning of the clause or sentence	3	3

5.1	Make sure readers can identify what each pronoun is referring to	2	2
5.2	Don't use *this*, *that*, *these*, and *those* as pronouns	3	2
5.3	Don't use *which* to refer to an entire clause	3	2

6.1	Don't use a telegraphic writing style	1	1
6.11	Make each sentence syntactically and semantically complete	2	1
6.3	Use *that* with verbs that take noun clauses as complements	2	2
6.4	Use *that* in relative clauses	2	2
6.5	Clarify which parts of a sentence are being joined by *and* or *or*	2	2
6.6	Revise past participles	2	2
6.7	Revise adjectives that follow nouns	2	2
6.8	Use *to* with indirect objects	2	3

(*continued*)

6.2	In a series of noun phrases, consider including an article in each noun phrase	3	2
6.9	Consider using *both . . . and* and *either . . . or*	3	3
6.10	Consider using *if . . . then*	3	3

7.1	Revise -ING words that follow and modify nouns	2	2
7.4	Punctuate -ING phrases correctly	2	2
7.5	Hyphenate -ING words in compound modifiers	2	2
7.6	Eliminate unnecessary -ING phrases and -ING clauses	2	2
7.2	Revise -ING words that follow certain verbs	2	3
7.3	Revise dangling -ING phrases	3	1
7.7	Revise ambiguous -ING + noun constructions	3	3
7.8	Revise ambiguous *to be* + -ING constructions	3	3

Prioritized for Machine Translation

As explained in Chapter 1, "Introduction to Global English," the Global English guidelines are primarily intended for documents that will be translated by human translators or that will be read in English by non-native speakers. The priority values that are assigned in the MT category are merely educated guesses about which guidelines would enable most machine-translation systems to translate sentences more accurately.

Before implementing machine-translation software, ask the software vendor to help you identify and prioritize the guidelines that would make your documentation most suitable for that particular software. Also consult sources such as Bernth and Gdaniec (2000), Roturier (2004), and O'Brien and Roturier (2007) to gain a better understanding of how to evaluate the effect of specific guidelines on MT output.

Even if you are using machine-translation software, you probably don't use it for all of the languages that your documentation is translated into. Therefore, the HT (human translation) and NN (non-native speakers) guidelines are still important.

In the following tables, the guidelines are sorted first by MT, and then by HT and NN.

Sorted in a Single Table

Number	Description	MT	HT	NN
2.1	Be logical, literal, and precise in your use of language	1	1	1
3.1	Limit the length of sentences	1	1	1
6.1	Don't use a telegraphic writing style	1	1	1
3.7	Consider defining, explaining, or revising noun phrases	1	1	2
6.11	Make each sentence syntactically and semantically complete	1	1	2
2.3	Don't add verb suffixes or prefixes to nouns, acronyms, initialisms, or conjunctions	1	1	3
2.2	Use nouns as nouns, verbs as verbs, and so on	1	2	2
2.4	Use standard verb complements	1	2	2
7.6	Eliminate unnecessary -ING phrases and -ING clauses	1	2	2
3.6	Limit your use of passive voice	1	2	3
3.8	Use complete sentences to introduce lists	1	2	3
2.5	Don't use transitive verbs intransitively, or vice versa	1	3	3
2.6	Use conventional word combinations and phrases	1	3	3
3.4	Keep phrasal verbs together	1	3	3
4.1	Place *only* and *not* immediately before whatever they are modifying	1	3	3
7.3	Revise dangling -ING phrases	2	1	3
3.3	Use a verb-centered writing style	2	2	2
3.11	Avoid ambiguous verb constructions	2	2	2
3.12	Write positively	2	2	2
4.2	Clarify what each prepositional phrase is modifying	2	2	2
4.3	Clarify what each relative clause is modifying	2	2	2
5.1	Make sure readers can identify what each pronoun is referring to	2	2	2
6.3	Use *that* with verbs that take noun clauses as complements	2	2	2
6.4	Use *that* in relative clauses	2	2	2
6.5	Clarify which parts of a sentence are being joined by *and* or *or*	2	2	2
6.6	Revise past participles	2	2	2

(*continued*)

6.7	Revise adjectives that follow nouns	2	2	2
7.1	Revise -ING words that follow and modify nouns	2	2	2
7.4	Punctuate -ING phrases correctly	2	2	2
7.5	Hyphenate -ING words in compound modifiers	2	2	2
4.4	Use *that* in restrictive relative clauses	2	2	3
4.6	Clarify ambiguous modification in conjoined noun phrases	2	2	3
5.2	Don't use *this*, *that*, *these*, and *those* as pronouns	2	2	3
5.3	Don't use *which* to refer to an entire clause	2	2	3
6.2	In a series of noun phrases, consider including an article in each noun phrase	2	2	3
3.10	Avoid unusual constructions	2	3	2
6.8	Use *to* with indirect objects	2	3	2
7.2	Revise -ING words that follow certain verbs	2	3	2
3.5	Use short, simple verb phrases	2	3	3
3.9	Avoid interrupting sentences	2	3	3
4.5	Consider moving anything that modifies a verb to the beginning of the clause or sentence	2	3	3
2.7	Don't use non-standard comparative and superlative adjectives	3	3	3
2.8	Use *the* only with definite nouns	3	3	3
2.9	Use singular and plural nouns correctly	3	3	3
3.2	Consider dividing shorter sentences	3	3	3
6.9	Consider using *both . . . and* and *either . . . or*	3	3	3
6.10	Consider using *if . . . then*	3	3	3
7.7	Revise ambiguous -ING + noun constructions	3	3	3
7.8	Revise ambiguous *to be* + -ING constructions	3	3	3

Grouped by Chapter

Number	Description	MT	HT	NN
2.1	Be logical, literal, and precise in your use of language	1	1	1
2.3	Don't add verb suffixes or prefixes to nouns, acronyms, initialisms, or conjunctions	1	1	3
2.2	Use nouns as nouns, verbs as verbs, and so on	1	2	2
2.4	Use standard verb complements	1	2	2
2.5	Don't use transitive verbs intransitively, or vice versa	1	3	3
2.6	Use conventional word combinations and phrases	1	3	3

(continued)

2.7	Don't use non-standard comparative and superlative adjectives	3	3	3
2.8	Use *the* only with definite nouns	3	3	3
2.9	Use singular and plural nouns correctly	3	3	3

3.1	Limit the length of sentences	1	1	1
3.7	Consider defining, explaining, or revising noun phrases	1	1	2
3.6	Limit your use of passive voice	1	2	3
3.8	Use complete sentences to introduce lists	1	2	3
3.4	Keep phrasal verbs together	1	3	3
3.3	Use a verb-centered writing style	2	2	2
3.11	Avoid ambiguous verb constructions	2	2	2
3.12	Write positively	2	2	2
3.10	Avoid unusual constructions	2	3	2
3.5	Use short, simple verb phrases	2	3	3
3.9	Avoid interrupting sentences	2	3	3
3.2	Consider dividing shorter sentences	3	3	3

4.1	Place *only* and *not* immediately before whatever they are modifying	1	3	3
4.2	Clarify what each prepositional phrase is modifying	2	2	2
4.3	Clarify what each relative clause is modifying	2	2	2
4.4	Use *that* in restrictive relative clauses	2	2	3
4.6	Clarify ambiguous modification in conjoined noun phrases	2	2	3
4.5	Consider moving anything that modifies a verb to the beginning of the clause or sentence	2	3	3

5.1	Make sure readers can identify what each pronoun is referring to	2	2	2
5.2	Don't use *this*, *that*, *these*, and *those* as pronouns	2	2	3
5.3	Don't use *which* to refer to an entire clause	2	2	3

6.1	Don't use a telegraphic writing style	1	1	1
6.11	Make each sentence syntactically and semantically complete	1	1	2

(continued)

6.3	Use *that* with verbs that take noun clauses as complements	2	2	2
6.4	Use *that* in relative clauses	2	2	2
6.5	Clarify which parts of a sentence are being joined by *and* or *or*	2	2	2
6.6	Revise past participles	2	2	2
6.7	Revise adjectives that follow nouns	2	2	2
6.2	In a series of noun phrases, consider including an article in each noun phrase	2	2	3
6.8	Use *to* with indirect objects	2	3	2
6.9	Consider using *both . . . and* and *either . . . or*	3	3	3
6.10	Consider using *if . . . then*	3	3	3

7.6	Eliminate unnecessary -ING phrases and -ING clauses	1	2	2
7.3	Revise dangling -ING phrases	2	1	3
7.1	Revise -ING words that follow and modify nouns	2	2	2
7.4	Punctuate -ING phrases correctly	2	2	2
7.5	Hyphenate -ING words in compound modifiers	2	2	2
7.2	Revise -ING words that follow certain verbs	2	3	2
7.7	Revise ambiguous -ING + noun constructions	3	3	3
7.8	Revise ambiguous *to be* + -ING constructions	3	3	3

Appendix C

Revising Incomplete Introductions to Unordered Lists

Introduction

The examples in this appendix are intended to help you follow guideline 3.8, "Use complete sentences to introduce lists." No matter what type of introductory phrase you have used, you will probably find a similar example here that shows you how to turn it into a complete sentence.

If you don't understand the grammatical terminology that is used in the headings, just skim through the examples to find one that is similar to the one that you're trying to revise.

Modal Verb Separated from Main Verbs

✗ In addition to invoking, managing, and scrolling windows, the windowing environment **can**

- customize windows
- manage libraries and files
- search text.

✓ In addition to invoking, managing, and scrolling windows, the windowing environment **can be used as follows**:

- to customize windows
- to manage libraries and files
- to search text

✗ When deploying Java applets, you **must**

- verify that your browser client has the necessary Java classes to run the applet
- verify that the Web server has the HTML files, the user-generated files, and any Java classes that it needs for running the applet
- select the type of HTML that you want to generate for the applet.

✓ When deploying Java applets, you **must do the following**:

- Verify that your browser client has the necessary Java classes to run the applet.
- Verify that the Web server has the HTML files, the user-generated files, and any Java classes that it needs for running the applet.
- Select the type of HTML that you want to generate for the applet.

Infinitive Marker *to* Separated from Infinitives

✗ To deploy a Java application, you need **to**

- create an executable
- set the PATH environment variable
- execute the application with either the `java` command or the `jre` command.

✓ To deploy a Java application, you need **to perform the following tasks**:

- Create an executable.
- Set the PATH environment variable.
- Execute the application with either the `java` command or the `jre` command.

Relative Pronoun Separated from the Rest of Some Relative Clauses

✗ JSP technology separates the user interface from the application logic, **which**

- enables the page designer to focus on writing the HTML that controls the page design
- enables the software developer to generate the dynamic-content portion of the page.

✓ JSP technology separates the user interface from the application logic. **As a result,** the page designer can focus on writing the HTML that controls the page design, and the software developer can focus on generating the dynamic-content portion of the page.

✗ InformationBeans are JavaBeans **that**

- access or process items on the server
- provide access to any data that SAS can access
- provide computational services such as statistical analysis, reporting, and quality control.

✓ InformationBeans are JavaBeans **that have the following additional capabilities**:

- They can access or process items on the server.
- They provide access to any data that SAS can access.
- They provide computational services such as statistical analysis, reporting, and quality control.

Preposition Separated from Its Objects

✗ In order to create or edit JavaServer pages, you should be familiar **with**

- HTML coding and Web page development
- Java programming, including the use of InformationBeans and TransformationBeans

✓ In order to create or edit JavaServer pages, you need the following skills:

- HTML coding
- Web page development
- Java programming, including the use of InformationBeans and TransformationBeans

Subject of Infinitives Separated from the Infinitives

✗ SAS/IntrNet software helps **organizations**

- share information with anyone who has a Web browser
- build and distribute applications via the Internet
- deploy sophisticated applications across the Web.

✓ SAS/IntrNet software gives you the following capabilities:

- You can share information with anyone who has a Web browser.
- You can build and distribute applications via the Internet.
- You can deploy sophisticated applications across the Web.

Subject Separated from Verbs

In this example the list was short enough that it was perfectly OK to turn it into a paragraph.

X ODS is especially helpful if **you**

- want to reuse your existing SAS programs
- generate reports that use static data.

✓ ODS is especially helpful if you want to reuse your existing SAS programs, or if you generate reports that use static data.

Verb Separated from Its Direct Objects

X XYZ software **provides**

- rapid, code-free creation of graphical reports
- easy access to enterprise-wide information over the Web
- a fully customizable document design.

✓ XYZ software **provides the following functionality**:

- rapid, code-free creation of graphical reports
- easy access to enterprise-wide information over the Web
- a fully customizable document design

✗ Examples of native drivers **include**:

✓ Here are some examples of native drivers:

✗ webDev features **include**:

- a visual drag-and-drop application builder
- comprehensive support for JavaBeans
- a GUI-based class editor.

✓ webDev software **includes the following features**:

- a visual drag-and-drop application builder
- comprehensive support for JavaBeans
- a GUI-based class editor

✗ In order to create and use JavaServer pages, you **must have**

- a Web server (such as Apache, Microsoft IIS, or Netscape Enterprise)
- an application server or a servlet engine that implements the Java Servlet API and JSP standards.

✓ In order to create and use JavaServer pages, you **must have the following**:

- a Web server (such as Apache, Microsoft IIS, or Netscape Enterprise)
- an application server or a servlet engine that implements the Java Servlet API and JSP standards

✗ Because Java applications cannot rely on Web browsers, they **require**

- access to the class, resource, and data files that the application uses
- a run-time environment (the Java virtual machine)
- an installation procedure that creates the necessary files, directories, and subdirectories.

✓ Because Java applications cannot rely on Web browsers, they **have the following requirements**:

- access to the class, resource, and data files that the application uses
- a run-time environment (the Java virtual machine)
- an installation procedure that creates the necessary files, directories, and subdirectories

Interrupted -ING Phrase

X In order to use formatting macros, you must be familiar with Base SAS software, **including**

- using the SAS windowing environment
- creating and submitting simple SAS programs
- using the SAS macro language.

✓ In order to use formatting macros, you must know how to do the following:

- use the SAS windowing environment
- create and submit simple SAS programs
- use the SAS macro language

Gerund Separated from Its Objects

X You can choose the appropriate scenario by **determining**

- where classes are stored (on the Web server or on the client machine)
- how classes are organized (individually or in a group archive file).

✓ You can choose the appropriate scenario by **answering two questions**:

- Are classes stored on the Web server or on the client machine?
- Are classes organized individually, or are they in a group archive file?

Appendix D

Improving Translatability and Readability with Syntactic Cues

Preface

A different version of this information was published as Kohl (1999). The syntactic cues procedure that was referred to in that article has been incorporated into several chapters of this book.

As its title indicates, Chapter 6, "Using Syntactic Cues," deals exclusively with syntactic cues. In addition, the following guidelines from other chapters could be classified as syntactic cues guidelines:

- 4.2.3, "If a prepositional phrase modifies a noun phrase, consider expanding it into a relative clause"
- 4.6, "Clarify ambiguous modification in conjoined noun phrases"
- 7.1, "Revise -ING words that follow and modify nouns"
- 7.2, "Revise -ING words that follow certain verbs"
- 7.3, "Revise dangling -ING phrases"
- 7.4, "Punctuate -ING phrases correctly"
- 7.5, "Hyphenate -ING words in compound modifiers"
- 7.7, "Revise ambiguous -ING + noun constructions"
- 8.3.1, "Use commas to prevent misreading"
- 8.3.2, "Use commas to separate main clauses"
- 8.3.3, "Consider using a comma before *because*"
- 8.3.4, "Consider using a comma before *such as*"
- 8.7.1, "Consider hyphenating noun phrases"
- 8.13.3, "Don't capitalize common nouns"
- 8.13.4, "When necessary, use capitalization to improve readability"

Introduction

In recent decades, many technical communicators have proposed guidelines for writing English-language documents that are intended for international audiences. These guidelines include the following:

- Minimum Word Strategy, in which illustrations are used in place of text (Gange and Lipton 1984; Hodgkinson and Hughes 1982; Hoffman 1998; Pearsall 1988; Twyman 1985; Vogt 1986)

- Controlled English and Simplified English (Dale and O'Rourke 1976; Gingras 1987; Hinson 1991; Kirkman, Snow, and Watson 1978; Kleinman 1982; Ogden 1932; Ogden 1942; Peterson 1990; Sakiey and Fry 1979; Strong 1983; Thomas et al. 1992; White 1981)
- terminology guidelines such as using consistent nomenclature and avoiding words that have multiple common meanings (Downs 1980; Megathlin and Langford 1991; Nadziejka 1991; Olsen 1984; Phillips 1991; Velte 1990)
- production guidelines such as planning for text expansion (because other languages are less succinct than English), separating text from illustrations, and taking different paper sizes into account (Hartshorn 1987; Holden 1980; Klein 1988; Matkowski and Coar 1983)
- cultural guidelines such as avoiding humor; avoiding culture-specific examples in text, photos, and symbols (Bosley 1996); being aware of culture-specific connotations of colors (Sanderlin 1988; Swenson 1988); and, more generally, being sensitive to cultural differences (Greenwood 1993; Hoft 1995).

Other guidelines focus on improving readability at the sentence, clause, and phrase level. Most of these are general readability principles such as using short sentences, using passive voice only when appropriate, keeping subjects and verbs close together, and avoiding long noun strings and nominalizations.

Sprinkled throughout the literature are guidelines that aim more specifically to make translators' jobs easier or to improve the readability of English-language documents for non-native speakers (Hunt and Kirkman, 1986; Buican, Hriscu, and Amador, 1993; and many others).

Like those last two types of guidelines, the syntactic cues guidelines focus on individual sentences, clauses, and phrases, and they take both translatability and readability into account. They are based on the linguistic analysis and classification of tens of thousands of sentences (primarily from software documentation) during the past twenty years, and they are supported by research from several disciplines.

Understanding the benefits and limitations of the syntactic cues guidelines—and learning to apply the guidelines appropriately— requires time and effort. But for many types of documentation, the time and effort are well spent.

What Are Syntactic Cues?

Syntactic cues are elements or aspects of language that help readers correctly analyze sentence structure, identify parts of speech, or both. For example, suffixes, articles, prepositions, auxiliary verbs, and word order enable us to make grammatical sense out of the following lines from Lewis Carroll's "Jabberwocky," even though the content words are nonsense:

> 'Twas brillig, and the slithy toves
> Did gyre and gimble in the wabe.

For example, we know that *toves* is a noun because it ends in *-s* and is preceded by the article *the*. We know that *slithy* is an adjective because it ends in *-y* (a typical adjectival suffix) and because it occurs between an article and a noun. *Gyre* and *gimble* must be verbs because of the presence of the auxiliary verb *did*.

Syntactic cues exist in all languages, though the specifics are of course different. In English and many other Western languages, the articles *a*, *an*, and *the* (or their equivalents) are significant syntactic cues. Languages such as Chinese and Japanese don't have articles, but they have other syntactic cues that English lacks.

In most cases, syntactic cues are not optional. We cannot arbitrarily omit suffixes or prefixes, nor can we omit articles, prepositions, or auxiliary verbs except under certain circumstances. However, this chapter focuses exclusively on syntactic cues that *are* optional in certain contexts. Only the following syntactic cues are discussed:

- *that*
- *that* + the verb *to be*
- the articles *a*, *an*, and *the*
- *to* (both as a preposition and as an infinitive marker)
- modal verbs such as *can*, *should*, and *might*
- auxiliary verbs such as *is/are/was/were*, *has/have/had*, and *has been/have been/had been/will have been*
- prepositions such as *by*, *for*, *with*, and *in*
- correlative pairs such as *either . . . or*, *both . . . and*, and *if . . . then*
- punctuation such as hyphens, commas, and parentheses
- pronoun or noun subjects

Kimball (1973) examines the role of several of these syntactic cues in sentence perception, and he explains their benefits in terms of transformational grammar. Since then, several technical communicators have touched on syntactic cues, though they don't refer to them by that name. For example, Weiss (1998) tells authors to use "optional words," and Jones (1996) suggests using hyphens to clarify noun strings. Kulik (1995) advises authors to refrain from omitting *that*, and Buican (1993) recommends avoiding ellipsis.

In spite of this good advice, the historical emphasis on conciseness in technical communication leads many technical writers and editors to routinely and deliberately eliminate syntactic cues from their documents. This is unfortunate because, as the rest of this article will show, the benefits of syntactic cues can be clearly demonstrated and are supported by research from many disciplines.

Benefits of Syntactic Cues

Inserting a syntactic cue is not always the best way to improve a problematic clause or sentence. However, syntactic cues often improve readability and translatability in one or more of the following ways:

- They enable readers, translators, and machine-translation systems to *analyze* sentence structure more quickly and accurately. In this respect, they are particularly beneficial for non-native speakers of English.
- They make it easier for readers to *predict* the structure of subsequent parts of a sentence. Facilitating prediction can increase reading speed.
- They eliminate ambiguities that might not be noticed by human translators and that can therefore result in mistranslation. Ambiguities also are likely to be mistranslated by machine-translation systems, and they often force human translators to seek clarification or to make educated guesses.

In contrast to many implementations of controlled English, the syntactic cues guidelines don't impose inordinate restrictions on vocabulary nor on the range of grammatical constructions that are permitted. When used with discretion, the syntactic cues guidelines don't result in language that sounds unnatural to native speakers of English.

Facilitating analysis

It is easy to see that inserting a syntactic cue can make it easier for all readers (both native speakers and non-native speakers of English)—and probably also for many machine-translation (MT) systems—to correctly analyze the structure of some sentences. For example, consider the following sentence:

1a After a process creates a resource, any process it starts inherits the resource identifiers.

Human readers, as well as MT systems, are likely to stumble on the main clause because two subjects, *process* and *it*, appear to be followed by two verbs, *starts* and *inherits*. This sequence is an apparent violation of normal word order in English sentences. It is much easier to recognize that the main clause contains an embedded relative clause when we insert the relative pronoun (syntactic cue) *that*:

1b After a process creates a resource, any process ***that*** *it starts inherits the resource identifiers.*

When a relative pronoun is the subject of a clause, you can often omit the relative pronoun plus a form of the verb *to be*. For example, in sentence 2b, the relative pronoun *that* and the verb *are* have been omitted. Sentence 2a is more syntactically explicit.

2a Programs ***that are*** *currently running in the system are indicated by icons in the lower part of the screen.*

2b Programs currently running in the system are indicated by icons in the lower part of the screen.

Non-native speakers who are not fluent in English have particular difficulty when *that* + *to be* is omitted. The participles that are left behind (such as *running* in the above example) can play so many different grammatical roles that they are inherently confusing to non-native speakers (Cohen et al. 1979; Berman 1984). In fact, participles are so problematic that *present* participles (the *-ing* verb forms) should be replaced by other constructions when possible. *Past* participles, which usually end in *-ed*, cannot be replaced so easily, but often they can be expanded into relative clauses. (See guideline 6.6.1, "Revise past participles that follow and modify nouns," for further discussion.) For example, you could replace sentence 3a with sentence 3b, with no loss of meaning and no change in emphasis.

3a DATAMAX continues ***processing*** *program statements after* ***repairing*** *the data set.*

3b DATAMAX continues ***to process*** *program statements after* ***it repairs*** *the data set.*

As sentences 3a and 3b illustrate, the syntactic cues approach is more than just inserting syntactic cues here or there. It often involves replacing an ambiguous or potentially confusing sentence constituent with something that is simpler or more syntactically explicit.

Facilitating prediction

The reading process has been characterized as a "psycholinguistic guessing game" (Goodman 1967). As we read, we are constantly predicting what's going to come next. Strong evidence for this claim is provided by "garden-path" sentences like the following:

4) *The cotton shirts are made from comes from Arizona.*

5) *Since Jay always jogs a mile doesn't seem that far to him.*

If we weren't continually analyzing or hypothesizing about the structures of these sentences as we read, then how could we explain the fact that these sentences "lead us down the garden path"? (That is, they don't turn out the way we expect them to.)

The importance of prediction in the reading process is supported by the fact that we can read a meaningful sentence aloud more rapidly than we can when the word order is jumbled or reversed (Smith 1988; Cziko 1978). Jumbling the words of course renders the sentence meaningless and makes prediction impossible. A study done in Sweden also found that reading speed increased when relative pronouns were included in Swedish (Platzach 1974).

Just and Carpenter (1987) state that because any single type of cue can be "weak, absent, or even misleading on occasion," readers rely on evidence from all available cues when they analyze the structure of a sentence (p. 145). One goal of the syntactic cues approach, then, is to ensure that readers have all the syntactic cues they need to immediately arrive at a correct analysis and prediction for each part of a sentence. For example, the following sentence leads many readers down the garden path:

6a *In Experiment 6 we were interested in the reading subjects spontaneously achieve for such a headline.*

To prevent *reading* from being misinterpreted as an adjective, we could insert the syntactic cue *that* after it:

6b *In Experiment 6 we were interested in the reading* ***that*** *subjects spontaneously achieve for such a headline.*

Alternatively, we might decide that a more drastic revision is desirable—one that uses the word *interpretation* in place of *reading*:

6c *In Experiment 6 we were interested in each subject's first* ***interpretation*** *of this type of headline.*

Resolving ambiguities

Sometimes syntactic cues do more than make sentences easier to analyze and comprehend: they often eliminate ambiguities that confuse native speakers as well as non-native speakers and translators. For example, when the conjunction *and* is used, it is not always clear which parts of the sentence are being conjoined.

7) *Semaphores enable an application to signal completion of certain tasks and* ***[to? | they?]*** *control access to resources that more than one process might need to use.*

In this case (as with many other "scope of conjunction" ambiguities), the difference in meaning between the two possible interpretations is slight. However, the ambiguity is an impediment to translation—especially if machine-translation software is being used. The simple act of inserting a syntactic cue (in this case, the infinitive marker *to*) resolves the ambiguity without making the sentence sound unnatural. Indeed, inserting *to* adds a pleasing parallelism to the sentence. Many authors occasionally insert syntactic cues for that reason, even if they are unaware of the other benefits of syntactic cues and of the full scope of the syntactic cues approach.

The articles *a*, *an*, and *the* can sometimes be used as syntactic cues to resolve the potential semantic ambiguity of the word *or*. For example, in sentence 8a, only an expert could know whether *exception* is a synonym for *hard error*, or whether they are two different things.

8a The system immediately terminates the program if a hard error or exception occurs.

If they are different things, then we can make that clear by inserting *an*, as in sentence 8b:

8b The system immediately terminates the program if a hard error or ***an*** *exception occurs.*

We can make it even clearer by using *either . . . or* along with the article *an*, as in sentence 8c.

8c The system immediately terminates the program if ***either*** *a hard error* ***or an*** *exception occurs.*

If *hard error* and *exception* are synonyms, then we could use parentheses to indicate their equivalence, if that convention is followed consistently within a document:

8d The system immediately terminates the program if a hard error **(or exception)** *occurs.*

Guideline 4.6, "Clarify ambiguous modification in conjoined noun phrases," and guideline 6.5, "Clarify which parts of a sentence are being joined by *and* or *or*," provide further explanations of the types of ambiguity that occur frequently in association with the conjunctions *and* and *or*.

Benefits of syntactic cues for non-native speakers of English

A huge volume of technical information is available only in English and must therefore be read in English by speakers of other languages. For example, documentation for highly technical products is not necessarily translated into all the languages of the countries where the products are sold. Many academic journals and reports of scientific research are published only in English, and a large percentage of the information on the World Wide Web is in English. Thus, non-native speakers of English constitute an important part of the audience for many documents.

Syntactic cues are particularly important for these readers. Unless they are extremely fluent, non-native speakers interpret individual words and syntactic structures more slowly than native speakers, and they are less able to anticipate the sequence of words (Macnamara 1970). Hatch, Polin, and Part (1974) found that first-language readers rely primarily on content words (nouns, verbs, adjectives, and adverbs), whereas second-language readers rely both on content words and on syntactically redundant function words (articles, conjunctions, and prepositions)—that is, on syntactic cues. Cziko (1978) found that "sensitivity to syntactic constraints develops before sensitivity to semantic and discourse constraints" (p. 485). He defines syntactic constraints as "constraints [that are] provided by the preceding words and [by] the syntactic rules of the language—for example, that the word *the* will most likely be followed by a noun." Thus, syntactic cues help non-native speakers to interpret syntactic structures more rapidly and to better anticipate the sequence of words.

Imperfect knowledge of English, as well as interference from their native languages, causes non-native readers to have particular problems with certain syntactic features of English. For example, as noted earlier, participles (verb forms that end in *-ing* or *-ed*) are problematic because they can fill so many different grammatical functions. Berman (1984) mentions several other sources of difficulty, including lack of relative pronouns in relative clauses, deletion of *that* plus the verb *to be* in modifiers that follow nouns, and omission of *that* following verbs that take noun clauses as complements. In this book, guideline 6.6, "Revise past participles," and guideline 7.1, "Revise -ING words that follow and modify nouns," suggest ways of expanding participles into explicit relative clauses (making them less ambiguous) or of eliminating participles completely. Guideline 6.3 addresses the use of *that* with verbs that take noun clauses as complements.

Swan, Smith, and Ur (2001) provide fascinating details about other difficulties that non-native speakers from dozens of language groups are likely to have with English.

Caveat Scriptor: Let the Writer Beware!

Remember: Inserting a syntactic cue is not always the best way to revise a sentence that is ambiguous or difficult for readers to parse. Sometimes a more drastic revision is appropriate. For example, for most readers, sentence 9a is a garden-path sentence:

9a *To create a table and load it with a subset of data, create a DBMS view, and view the subsetted data, follow these steps:*

You could prevent misreading by inserting the syntactic cue *to*, as follows:

9b *To create a table and load it with a subset of data,* **to** *create a DBMS view, and* **to** *view the subsetted data, follow these steps:*

However, a more drastic revision, possibly putting each subtask under a separate heading, would almost certainly be in order.

Sometimes inserting a syntactic cue is unnecessary and makes the sentence sound unnatural. For example, in sentence 10, the infinitives *create* and *manage* are joined by *and.*

10) *A file system enables applications to create and* **[to]** *manage file objects.*

In this sentence it is not necessary to insert the syntactic cue *to* in front of *manage*, because *manage* immediately follows *and.* There are no intervening words that could lead to any confusion, ambiguity, or misreading.

At times, inserting a syntactic cue is not only unnecessary but would actually distort the meaning of the sentence. Consider the following sentence:

11a *You can use the REDO command to recall the statements that produced the errors,* **[to]** *make the corrections based on the log information, and* **[to]** *resubmit the program.*

If you insert the infinitive markers (*to*), then you are saying that the REDO command is used not only to recall statements, but also to make corrections and to resubmit the program. Even someone who is not familiar with the subject matter can guess that this interpretation is incorrect. Sentence 11b is an accurate revision of 11a:

11b *You can use the REDO command to recall the statements that produced the errors. Next, use the information in the log to correct the statements, and then resubmit the program.*

Sometimes a sentence must be rearranged or restructured slightly in order to insert a syntactic cue:

12a Before printing a monetary value, you usually assign it a format.

12b Before printing a monetary value, you usually assign a format ***to*** *it.*

In other cases it might be appropriate to restructure a sentence *instead of* using a syntactic cue. For example, 13b uses the syntactic cue *that*, but 13c simplifies the sentence structure by eliminating the relative clause altogether and using *which* as an adjective.

13a The software automatically determines the type of file it is reading and reads the file accordingly.

13b The software automatically determines the type of file ***that*** *it is reading and reads the file accordingly.*

13c The software automatically determines ***which*** *type of file it is reading and reads the file accordingly.*

Often the best solution is to restructure a sentence *in addition to* using syntactic cues:

14a You can develop an application ***using*** *Appli-Pro that communicates with a client session* ***using*** *TCP/IP sockets.*

14b You can ***use*** *Appli-Pro to develop an application* ***that uses*** *TCP/IP sockets to communicate with a client session.*

Anyone who inserts syntactic cues without considering whether they are really necessary or whether other, better revisions are possible is not following the spirit of the Global English guidelines.

Considerations Regarding the Use of Syntactic Cues

In addition to the references that have already been cited, a considerable amount of other research from the fields of psycholinguistics and reading behavior supports the use of syntactic cues. Much of this research involves methodologies and theories that require more careful, detailed discussion than is possible in this chapter. However, some particularly interesting and noteworthy findings are summarized below. (For additional reading, see Martin and Roberts 1966; Fodor and Garrett 1967; Hakes and Cairns 1970; Hakes and Foss 1970; Hakes 1972; Dawkins 1975; Hakes, Evans, and Brannon 1976; Huggins and Adams 1980; Frazier and Rayner 1982; Clifton, Frazier, and Connine 1984; Frazier 1988; and Garrett 1990.)

There are different degrees of ambiguity and of sensitivity to ambiguity

In actual practice, authors seldom lead their readers "down the garden path" in ways that are startling or obvious, as in sentences 4 and 5, above. However, the fact that a sentence is ambiguous can easily go unnoticed by one person (the author), yet be painfully obvious to another person (a translator or a novice user, for example).

According to Schluroff et al. (1986), we encounter ambiguous words and sentences so often in our everyday use of language that we usually don't even notice the ambiguities. However, there are different degrees of ambiguity, and "ambiguous sentences usually have a more or less pronounced bias toward one of their meanings, which means that one of the meanings is usually preferred over the other, either in general, in a given context, or to an individual" (p. 324). For example, the newspaper headline "Girl, 13, Turns in Parents for Marijuana, Cocaine" is more likely to be interpreted as meaning "Girl, 13, Turns in Parents for *Using* Marijuana and Cocaine" than as "Girl, 13, Turns in Parents in Exchange for *a Reward of* Marijuana and Cocaine" (Perfetti et al., 1987, p. 692). The more highly biased an ambiguous sentence is toward one of its meanings, the more likely it is that the ambiguity (that is, the alternate interpretation) will go unnoticed.

Just as phrases or sentences can have different degrees of ambiguity, so the degree of sensitivity to ambiguity varies widely among individuals. Perfetti et al. (1987) report that "two [out of 48] subjects noticed both meanings for 22 of the 25 ambiguous headlines they read," while "at the other extreme were two subjects who, only with probing, noticed ambiguities in 2 of the 25 headlines, and one who noticed no ambiguities at all" (p. 705). Similarly, in a test of "grammatical sensitivity" administered by Schluroff et al. (1986), the scores ranged from 18 to 33.5 out of a possible 35 (p. 329).

This research suggests that translators and other readers of technical information might arrive at the unintended meaning of an ambiguous phrase or sentence without even realizing that the alternate, intended interpretation is possible. That possibility underscores the importance of identifying all of the contexts in which syntactic cues can help authors and editors eliminate ambiguity from technical information.

Context does not prevent misreading

One objection that subject-matter experts and even technical writers often raise when a garden-path sentence or other ambiguity in their writing is pointed out to them is that "the meaning is clear from the context." This objection raises the question of whether contextual knowledge actually comes into play early enough to prevent faulty analysis by the parser.

Ferreira and Clifton (1986) found that readers initially compute an incorrect syntactic analysis of certain sentences even when the preceding context provides information which, in principle, could have been used to analyze the sentence correctly. Even readers who are familiar with the subject matter of a text will occasionally analyze sentences incorrectly when too few syntactic cues have been provided. They might backtrack and correct their faulty analysis fairly quickly, but "it is costly in terms of perceptual complexity ever to have to go back to reorganize the constituents of [a] phrase" (Kimball 1973, p. 37).

Rayner, Carlson, and Frazier (1983) similarly found that "the relative plausibility of an event described by a sentence did not influence the initial parsing strategy, as evidenced by eye fixation times" (cited in Perfetti et al., 1987, p. 693). Perfetti et al. used newspaper headlines rather than sentences in their study, reasoning that the "syntactic impoverishment" of newspaper headlines would make it even more likely that plausibility would influence parsing strategy. They found that pragmatic information was indeed used by readers to interpret headlines, but that it was not brought in quickly enough to override syntactic processes, "even when it would be advantageous to do so" (p. 706).

Perfetti et al. (1987) also found that ambiguous headlines "take longer to comprehend even when preceded by context" (p. 701).

To sum up, it is clear that subject-matter knowledge doesn't prevent readers from being led down the garden path and from having to reanalyze ambiguous sentences. Hence, those syntactic cues that reduce the likelihood of ambiguity can be beneficial to expert readers and non-experts alike.

There is only one difference between expert and non-expert readers in this regard: after the initial, faulty analysis, experts are more likely to have the necessary contextual knowledge to reanalyze the sentence correctly.

The reading process differs according to purpose

As Alderson (1984) points out, the reading process differs depending on whether the reader's goal is global comprehension or local comprehension. For skimming—for "getting the gist of a text"—Berman (1984) says that syntax (and hence, syntactic cues) "may not be all that crucial." However, if acquiring specific information accurately and in detail is important, then "exact appreciation of [the] syntactic components of each sentence remains an important aim" (p. 146).

If you produce documents that are not likely to be read word-for-word, then adding syntactic cues might not be worthwhile. However, for most documents, readers skim only until they find the relevant section of the document. Thereafter, they typically read for detailed, local information, and syntactic cues again become important.

For some types of texts, syntactic cues might not be very helpful

If you are already using a fairly restrictive form of controlled English, in which sentences are short, terminology is tightly controlled, and sentence structures are already greatly simplified, then syntactic cues probably will not significantly improve the readability or translatability of your text.

At the other extreme, syntactic cues won't do much to help a document that contains serious stylistic or organizational problems or whose content is inappropriate for its audience. There is little value in inserting syntactic cues until after the more fundamental problems have been addressed.

Integrating Syntactic Cues into Your Documentation Processes

Overcoming concerns about word counts

The syntactic cues guidelines contradict the training of many authors who have been taught for years to eliminate every unnecessary *that*, to use punctuation sparingly, and, in general, to strive for brevity. In addition, management might be concerned about syntactic cues adding to the costs of translation and publication because of increased word counts.

If these concerns are an issue in your workplace, then show your colleagues examples from your own organization's documents of sentences that would be ambiguous or difficult to translate without syntactic cues. Count the words in the "before" and "after" versions. (In most cases, the increase in word count is negligible.) Be sure to note improvements that you made in the text as a result of systematically looking at contexts in which syntactic cues might be useful, even if you chose to use a different revision strategy.

In most cases, syntactic cues facilitate translation by reducing the number of ambiguities that translators are forced to resolve. Because resolving an ambiguity often means making time-consuming, costly, and frustrating attempts to contact a particular subject-matter expert, translators are sometimes forced to guess at which meaning was intended. Alternatively, they might settle for equally ambiguous constructions in the target languages, if such constructions exist. If a document is translated into 10 other languages, then 10 translators must deal with each ambiguity in the source.

As Hunt and Kirkman (1986) have pointed out, "any supposed gain from saving space by omitting [syntactic cues] is usually outweighed by the extra decoding processes [that are] forced on readers" (p. 155). Readers are far better served when authors reduce word counts by other means instead.[1]

Working with your translators

If your organization is receptive to the syntactic cues guidelines and to focusing more on translatability in general, then be sure to ask your translators or localization coordinator to help you decide which of the syntactic cues guidelines (and which of the other Global English guidelines) to focus on. Some syntactic cues are more important for Western European languages than for Asian languages, or vice versa. Some are important if you anticipate using machine-translation software, but are less important for human readers. Also refer to Appendix B, "Prioritizing the Global English Guidelines," for guidance on how to prioritize the guidelines.

Your translators might also suggest other ways of conveying information that would reduce the need for *some* syntactic cues. For example, in Chapter 1, the "Insert Explanations for Translators" topic suggests inserting comments to explain how translators should interpret certain ambiguous constructions.

These comments can be an excellent solution in cases where your primary audience might find particular syntactic cues intrusive. For example, in sentence 15a it might not be clear to a translator—and it would certainly be ambiguous to an MT system—that MDDB is an adjective (modifying *registration*) rather than a noun.

15a The GETMDDBINFO function queries the metabase to retrieve information about the MDDB or table registration.

If MDDB is a noun, then it would be better to insert syntactic cues in the sentence as follows:

15b The GETMDDBINFO function queries the metabase to retrieve information about the MDDB or ***about the*** *table registration.*

If MDDB is an adjective, the only way of making that fact clear to translators (including machine-translation systems) would be to repeat the word *registration*:

15c The GETMMDBINFO function queries the metabase to retrieve information about the MDDB ***registration*** *or the table registration.*

---

[1] See Chapter 1, "Introduction to Global English," and Appendix A, "Examples of Content Reduction," for a discussion and examples of content reduction.

If you feel that 15a would be clear to your intended audience and that 15b and 15c are stylistically unacceptable, then you could provide the appropriate expanded version of the sentence only to translators. Many publishing tools enable you to embed conditional text, which appears in one view of the information (in this case, the translators' view), but not in the production version. For example, if you are using XML or SGML, the sentence might appear as follows, where <translationNote> indicates a comment that appears only in the translators' view of the text:

15d *The GETMMDBINFO function queries the metabase to retrieve information about the MDDB and* `<translationNote>about the</translationNote>` *table registration.*

In any case, your translators will probably be happy to help you decide which types of information to provide and how to provide it.

Conclusion

The syntactic cues guidelines help fill a significant gap in the "writing for international audiences" literature. These guidelines are solidly supported by research, and they have been well received by translators, localization managers, and non-native speakers of English.

A thorough understanding of syntactic cues can help many authors take a giant step toward a writing style that is eminently suitable for a global audience.

Glossary

acronym

an abbreviation that is formed from the first letter or letters of a group of words. Acronyms are pronounced like words, as in NATO (NAY-toh). By contrast, initialisms are pronounced by saying the name of each letter, as in RCA (R-C-A).

clause

a unit of sentence structure that includes a verb and usually a subject. See also *main clause*, *subordinate clause*.

cognitive dissonance

psychological discomfort that arises when an individual encounters something that is contrary to his or her beliefs, values, or expectations.

collocation

two or more words that are used in close proximity to each other more frequently than would be expected by chance. For example, the verbs *perform* and *schedule* are used more frequently with the noun *surgery* than other verbs.

colloquialism

an expression that is not appropriate for formal speech or writing. Example: *One way of avoiding* spam *is to <u>munge</u> your e-mail address.*

complement

See *verb complement.*

controlled-authoring software

software that parses texts and brings style errors, grammar errors, and terminology errors to the user's attention.

controlled English

any of several versions of controlled language that have been developed for English. See also *controlled language*.

controlled language

a subset of a natural language in which syntax, style, and terminology are restricted. Controlled languages are used to facilitate translation, to make the language easier for non-native speakers to understand, or both.

dangling participle

a participial phrase in which the subject of the participle is not the same as the subject of the main clauses. For example, the following sentence begins with a dangling participle: *Driving through the tunnel, an accident brought traffic to a standstill. An accident* is not the subject of *Driving*.

deprecate

to designate as incorrect or undesirable.

deprecated term

a term whose use is incorrect or undesirable.

determiner

a word that precedes a noun and that either quantifies or helps to identify the noun. In English, determiners include the following:

- articles (*a*, *an*, and *the*)
- numbers
- quantifiers (*many*, *much*, *some*, *several*, *a few*, and so on)
- demonstrative pronouns (*this*, *that*, *these*, and *those*)
- possessive pronouns (*my*, *your*, *their*, and so on).

ellipsis

in a sentence or phrase, the omission of one or more words that the author assumes are understood by the reader, but without which a construction is syntactically or semantically incomplete.

finite verb

a form of a verb that is inflected for person and tense. For example, in *I go, she goes, we went, they have gone*, the verb forms and endings change depending on the subject and the tense. See also *infinitive*.

gerund

an -ING word that is used as a noun. For example, in *Gardening is hard work*, and *Suzy hates swimming*, the -ING words are gerunds. In this book, the term *-ING word* encompasses gerunds and present participles. See also *-ING word*, *present participle*.

head noun

the main noun in a noun phrase. The head noun can be preceded by one or more articles (*a*, *an*, or *the*), adjectives, adverbs, determiners, or other nouns.

idiom

a group of words whose meaning is not derived from the literal meanings of the individual words.

infinitive

a form of a verb that is not inflected or conjugated. Infinitives are usually preceded by *to*, as in *El Niño conditions started to develop in the Central Pacific Ocean during August*. See also *finite verb*.

-ING clause

See "-ING Clauses" in Chapter 7, "Clarifying -ING Words."

-ING phrase

See "-ING Phrases" in Chapter 7, "Clarifying -ING Words."

-ING word

a word that is formed by adding the suffix *-ing* to the root form of a verb. For example, *doing*, *excavating*, and *restoring* are -ING words. -ING words are discussed in detail in Chapter 7, "Clarifying -ING Words." See also *gerund*, *present participle*.

initialism

an abbreviation that is formed from the first letter or letters of a group of words. Initialisms are pronounced by saying the name of each letter, as in RCA (R-C-A). By contrast, acronyms are pronounced like words, as in NATO (NAY-toh).

intransitive verb

a verb that, in standard English, is not used with a direct object.

language technology

any type of software that helps solve problems of communicating via human language. Speech recognition, controlled authoring, and computer-assisted translation are examples of language technologies.

localization

the process of adapting products or services for a particular geographical region or market. Translation is a large part of the localization process.

machine translation

the use of software to translate texts from one language, such as English, to one or more other languages, such as French or Japanese.

main clause

a clause that can stand alone as a complete sentence. For example, in the following sentence, the underlined clause is a main clause: *While running through the woods, I tripped and sprained my ankle*. See also *clause, subordinate clause*.

metaphor

a term or expression that is used in a non-literal sense in order to suggest a similarity.

MT

See *machine translation*.

non-native speaker of English

someone for whom English is a second language. For example, someone who grew up speaking Japanese but learned English later on is a non-native speaker of English. Non-native speakers usually are less fluent than native speakers of a language.

non-restrictive relative clause

one of two types of relative clauses. A non-restrictive relative clause provides non-essential information about the noun that it modifies. For example, the relative clause in the following sentence is non-restrictive: *Butterfly taxa, which have a low risk of extinction, are listed in the Least Vulnerable column*. See also *relative clause, restrictive relative clause*.

noun clause

a clause that can fill the same grammatical role as a noun phrase. For example, a noun clause can be a direct object, as in *Please ensure that your seat backs are locked in their full upright positions*. A noun clause can be a subject, as in *Whatever you dream can become a reality*. And a noun clause can be the object of a preposition, as in *I heard about what you said about me*.

noun phrase

a phrase that has a noun or a pronoun as its head. For example, *a beautiful sunset* is a noun phrase in which *sunset* is the head. A noun phrase can also consist of a single noun or pronoun without any articles or modifiers.

orthography

how terms are written—as one word or two, hyphenated or not, capitalized or not, and so on. By contrast, *spelling* refers to the sequence of non-blank characters that are used, without regard to capitalization or hyphenation.

parallelism

a writing technique in which the same or similar grammatical constructions are used for related words, phrases, or clauses.

participial phrase

a group of words that consists of either a past participle or a present participle, plus a number of other words. Here is an example of a past participial phrase: *The CD flew across the room, ejected by the drive in an apparent rejection of gangsta music*. And here is an example of a present participial phrase: *The piston moved downward, compressing the mixture of air and fuel in the crankcase*. See also *past participle, present participle.*

participle

See *past participle, present participle.*

past participle

the form of a verb that is used in all perfect tenses (as in *I have seen the light*) and to form the passive voice (as in *That remains to be seen*). Past participles are also used as adjectives (as in *a fully developed implementation plan*) and to introduce participial phrases (as in *The CD flew across the room, ejected by the drive in an apparent rejection of gangsta music*).

phrase

a word or group of words that form a syntactic unit and that have a single grammatical function. For example, English has noun phrases, verb phrases, adjective phrases, adverb phrases, prepositional phrases, and so on.

post-editing

the task of improving the quality of machine-translation output by correcting errors in word choice, grammar, and style. Post-editing is performed by human linguists, translators, or editors.

predicate adjective

an adjective that follows the main verb in a clause and that also follows the noun that it is modifying. For example, in *The night was dark and stormy*, *dark* and *stormy* are predicate adjectives that modify *night*. In *The dark night was stormy*, only *stormy* is a predicate adjective. In *It was a dark and stormy night*, neither is a predicate adjective.

present participle

a word that is derived from a verb and that ends in *-ing*. Present participles are used as follows: 1) to form the progressive aspect of verbs (as in *Sean will be running in the Boston Marathon*); 2) as adjectives (as in *a charging rhinoceros*); 3) to introduce present participial phrases (as in *The piston moved downward, compressing the mixture of air and fuel in the crankcase*). In this book, the term -ING word encompasses present participles and gerunds. See also *-ING word*, *gerund*.

progressive verb form

a form of a verb that indicates that an action is or was in progress at the time that a clause or sentence focuses on. Progressive tenses are formed by a form of the verb *to be* plus the –ING form of another verb, as in *Shirley was unloading her groceries when a thief stole her wallet from the grocery cart*.

referent

the noun that a pronoun such as *it* or *they* refers to.

relative clause

a clause that begins with a relative pronoun (*that*, *which*, *who*, or *whom*) and that modifies a preceding noun phrase. However, as explained in guideline 4.3, "Clarify what each relative clause is modifying," a relative clause doesn't necessarily modify the *closest* preceding noun phrase.

restrictive relative clause

one of two types of relative clauses. A restrictive relative clause provides information that is necessary in order for the reader or listener to identify which specific instance of a noun is being referred to. For example, the relative clause in the following sentence is restrictive: *(Only) butterfly taxa that have a low risk of extinction are listed in the Least Vulnerable column*. See also *relative clause*, *non-restrictive relative clause*.

SAS

an acronym for SAS Institute, Inc. SAS is a large, privately owned software company whose headquarters is in Cary, North Carolina. Because SAS is an acronym rather than an initialism, it is pronounced "sass." In some of the example sentences in this book, SAS refers to SAS software.

scope of conjunction

the parts of a sentence that are joined by *and* or *or*. The scope of these conjunctions is often ambiguous. See guideline 6.5, "Clarify which parts of a sentence are being joined by *and* or *or*," for examples.

source language

the language in which a text was originally written. See also *target language*.

subordinate clause

a clause that cannot stand alone as a complete sentence. For example, in the following sentence, the underlined clause is a subordinate clause: *While running through the woods, I tripped and sprained my ankle*. See also *clause*, *main clause*.

syntactic cue

an element or aspect of language that helps readers to identify parts of speech or to analyze sentence structure correctly, but which can be omitted in some contexts without making a clause or sentence completely incomprehensible or ungrammatical. For example, suffixes, articles, prepositions, and auxiliary verbs, are syntactic cues.

target language

the language into which a translated text has been or will be translated. See also *source language*.

TM software

See *translation memory software*.

transitive verb

a verb that, in standard English, is used with a direct object.

transitivity

See *transitive verb*, *intransitive verb*.

translation memory software

software that stores matching segments of source-language text and (translated) target-language text in a database for future reuse. When a new or updated document is processed by the software, any segments that are identical or similar to previously translated segments are presented to the translator. The translator then decides whether to reuse, modify, or disregard the previous translations.

translation segment

a unit of translation that is used by translation memory software. Translation segments are delineated by punctuation marks (typically, periods, semicolons, colons, question marks, and exclamation points) and by formatting characters (for example, paragraph delimiters, tabs, and carriage returns). As a result, translation segments can be complete sentences, titles, headings, list items, the contents of table cells, and so on.

verb complement

a grammatical construction such as a direct object, indirect object, prepositional phrase, or infinitive phrase that helps to complete the action or idea in the predicate of a sentence. See guideline 2.4, "Use standard verb complements," for examples.

Bibliography

Adriaens, Geert, and Dirk Schreurs. 1992. "From COGRAM to ALCOGRAM: Toward a Controlled English Grammar Checker." *Proceedings of the Fourteenth International Conference on Computational Linguistics* (COLING-92), August 23-28, 1992. Nantes, France, pp. 595-601. Available at http://acl.ldc.upenn.edu/C/C92/C92-2090.pdf.

AECMA. 1995. *AECMA Simplified English: A Guide for the Preparation of Aircraft Maintenance Documentation in the International Aerospace Maintenance Language.* Brussels: European Association of Aerospace Industries.

Akis, Jennifer Wells, and William R. Sisson. 2002. "Improving Translatability: A Case Study at Sun Microsystems, Inc." *The Globalization Insider* Issue 4.5 (December). Available at http://www.lisa.org/globalizationinsider/2002/12/improving_trans.html.

Alderson, J. Charles. 1984. "Reading in a Foreign Language: A Reading Problem or a Language Problem?" *Reading in a Foreign Language*, edited by J. Charles Alderson and A. H. Urquhart, 1-24. London: Longman.

Allen, Jeffrey. 1999. "Adapting the Concept of 'Translation Memory' to 'Authoring Memory' for a Controlled Language Writing Environment." *Translating and the Computer 21 Conference Proceedings.* London: Aslib. Also available at http://www.transref.org/default.asp?docsrc=/u-articles/allen2.asp.

Allen, Jeffrey, and Kathleen Barthe. 2004. "Introductory Overview of Controlled Languages." Invited talk presented at the Society for Technical Communication meeting of the Paris, France chapter. April 2, 2004. Available from http://www.geocities.com/controlledlanguage.

Allen, Jeffrey. 2005. "An Introduction to Using MT Software." In the special supplement "Guide to Translation" of *MultiLingual Computing & Technology* 16.1: 8-12.

Altanero, Tim. 2000. "Translation: MT and TM." *Intercom* 48.5 (May): 24-26.

Amador, Mable, and Yvonne Keller. 2002. *International English Manual.* Los Alamos: Los Alamos National Laboratory. Available at http://www.bodeuxinternational.com/InternationalEnglishManual.pdf.

Antonopoulou, Katerina. 1998. "Resolving Ambiguities in SYSTRAN." *European Commission. Translation Services*, Vol. 98. Available at http://ec.europa.eu/translation/reading/articles/pdf/1998_01_tt_antonopoulou.pdf.

Arnold, Douglas J., et al. 1994. *Machine Translation: An Introductory Guide.* Manchester: NCC Blackwell. Available at http://www.essex.ac.uk/linguistics/clmt/MTbook/.

Bailie, Rahel Anne, and Jerome Ryckborst. 2002. "Reaching Global Audiences: Doing More with Less." *Intercom* 49.6 (June): 17-21.

Benson, Morton, Evelyn Benson, and Robert Ilson. 1997. *The BBI Dictionary of English Word Combinations,* Revised edition. Amsterdam/Philadelphia: John Benjamins.

Berman, Ruth A. 1984. "Syntactic Components of the Foreign Language Reading Process." In *Reading in a Foreign Language*, edited by J. Charles. Alderson and A.H. Urquhart, 139-159. London: Longman.

Bernth, Arendse. 1998a. "EasyEnglish: Preprocessing for MT." *Proceedings of the Second International Workshop on Controlled Language Applications* (CLAW98). Pittsburgh, Pennsylvania: Language Technologies Institute, Carnegie Mellon University, May 21-22, 1998, pp. 30-41.

Bernth, Arendse. 1998b. "Panel Discussion: Standardization." *Proceedings of the Second International Workshop on Controlled Language Applications* (CLAW98). Pittsburgh, Pennsylvania: Language Technologies Institute, Carnegie Mellon University, May 21-22, 1998, pp. 192-193.

Bernth, Arendse, and Claudia Gdaniec. 2000. "MTranslatability" (materials from a workshop on writing for machine translation, presented at the Association for Machine Translation in the Americas conference). Available at http://www.isi.edu/natural-language/organizations/amta/sig-mtranslatability-tutorial.htm.

Bernth, Arendse, and Claudia Gdaniec. 2001. "MTranslatability." *Machine Translation* 16.3 (September): 175-218.

Bosley, Deborah S. 1996. "International Graphics: A Search for Neutral Territory." *Intercom* 43.7: 4-7.

Buican, Ileana, Vivi Hriscu, and Mable Amador. 1993. "Using International English to Prepare Technical Text for Translation." *Proceedings of the 1993 International Professional Communication Conference.* 33-35.

Carté, Penny, and Chris Fox. 2004. *Bridging the Culture Gap: A Practical Guide to International Business Communication.* London: Kogan Page.

Celce-Murcia, Marianne, and Diane Larsen-Freeman. 1999. *The Grammar Book: An ESL/EFL Teacher's Course.* 2nd edition. Boston: Heinle & Heinle.

The Chicago Manual of Style, 15th ed. 2003. Chicago: University of Chicago Press.

Childress, Mark D. 2007. "Terminology work saves more than it costs." *MultiLingual* 18.3 (April/May): 43-46.

Clark, Bob. 2002. "E-term + TM = AM: Can translation memory and authoring memory go hand-in-hand?" *Language International* 14.4 (August): 21-25.

Clifton, C., Frazier, L., and Connine, C. 1984. "Lexical expectations in sentence comprehension." *Journal of Verbal Learning and Verbal Behavior* 23.6 (December): 696–708.

Close, R.A. 1975. *A Reference Grammar for Students of English.* London: Longman.

Cohen, Andrew, Hilary Glasman, Phyllis R. Rosenbaum-Cohen, Jonathan Ferrara, and Jonathan Fine. 1979. "Reading English for Specialized Purposes: Discourse Analysis and the Use of Student Informants." *TESOL Quarterly* 13.4 (December): 551-564.

Cziko, Gary A. 1978. "Differences in First- and Second-Language Reading: The Use of Syntactic, Semantic and Discourse Constraints." *Canadian Modern Language Review* 34.3 (February): 473-489.

Dale, Edgar, and Joseph O'Rourke. 1976. *The Living Word Vocabulary:* the words we know: a national vocabulary inventory. Elgin, IL: Dome.

Dawkins, John. 1975. *Syntax and Readability*. Newark, DE: International Reading Association.

Dillinger, Mike, and Arle Lommel. 2004. *LISA Best Practice Guide: Implementing Machine Translation*. Geneva: Localization Industry Standards Association. Available at http://www.lisa.org/Best-Practice-Guides.467.0.html.

Downs, Linn Hedwig. 1980. "So You're Multinational: How to Prepare Your Manuscripts for Translation." *Proceedings of the 27th International Technical Communication Conference*, W47-W49. Washington, D.C.: Society for Technical Communication.

Dunne, Keiran J. 2007. "Terminology: Ignore it at your peril." *MultiLingual* 18.3 (April/May): 32-38.

Farrington, Gordon. 1996. "AECMA Simplified English: An Overview of the International Aerospace Maintenance Language." *Proceedings of the First International Workshop on Controlled Language Applications* (CLAW96). Leuven, Belgium: Katholieke Universiteit Leuven Centre for Computational Linguistics, March 26-27, 1996, pp. 1-21.

Fenstermacher, Hans. 2006. "The Looming Crisis of Content." *ClientSideNews Magazine* 6.8 (August): 24-26. Available at http://www.translations.com/resources/downloads/editorials/Aug06CSN.pdf.

Ferreira, Fernanda, and Charles Clifton. 1986. "The Independence of Syntactic Processing." *Journal of Memory and Language* 25.3 (June): 348-368.

Fidura, Christie. 2007. "The benefits of managing terminology with tools." *MultiLingual* 18.3 (April/May): 39-41.

Fodor, J.A., and M. Garrett. 1967. "Some Syntactic Determinants of Sentential Complexity." *Journal of Perception and Psychophysics* 2: 289-296.

Frazier, Lyn. 1988. "The Study of Linguistic Complexity." In *Linguistic Complexity and Text Comprehension: Readability Issues Reconsidered*, edited by Alice Davison and Georgia M. Green. Hillsdale, NJ: Lawrence Erlbaum Associates. 193-221.

Frazier, L. and K. Rayner. 1982. "Making and Correcting Errors during Sentence Comprehension: Eye Movements in the Analysis of Structurally Ambiguous Sentences." *Cognitive Psychology* 14: 178-210.

Fuchs, Norbert E., and Rolf Schwitter. 1996. "Attempto Controlled English (ACE)." *Proceedings of the First International Workshop on Controlled Language Applications* (CLAW96). Leuven, Belgium: Katholieke Universiteit Leuven Centre for Computational Linguistics, March 26-27, 1996, pp. 124-136.

Fuchs, Norbert E., Uta Schwertel, Rolf Schwitter. 1999. *Attempto Controlled English (ACE) Language Manual*, Version 3.0. Institut für Informatik der Universität Zürich, Nr. 99-03. August 1999. Available at ftp://ftp.ifi.unizh.ch/pub/techreports/TR-99/ifi-99.03.pdf.

Gange, Charles, and Amy Lipton. 1984. "Word-free Setup Instructions: Stepping into the World of Complex Products." *Technical Communication* 31.3 (3rd Quarter): 17-19.

Garrett, Merrill F. 1990. "Sentence Processing." In *An Invitation to Cognitive Science*, vol.1 *Language*, edited by Daniel N. Osherson and Howard Lasnik. Cambridge, MA: MIT Press. 133-175.

Gingras, Becky. 1987. "Simplified English in Maintenance Manuals." *Technical Communication* 34.1 (February): 24-28.

Goodman, K.S. 1967. "Reading: A Psycholinguistic Guessing Game." *Journal of the Reading Specialist* 6: 126-135.

Goyvaerts, Patrick. 1996. "Controlled English: Curse or Blessing? A User's Perspective." *Proceedings of the First International Workshop on Controlled Language Applications* (CLAW96). Leuven, Belgium: Katholieke Universiteit Leuven Centre for Computational Linguistics, March 26-27, 1996, pp. 137-142.

Graefe, Richard. 2006. "State of Technical Editing." Post to the STCTESIG-L mailing list, September 27, 2006. Available at http://mailman.stc.org/pipermail/stctesig-l/2006-September/000789.html.

Greenwood, Timothy G. 1993. "International Cultural Differences in Software." *Digital Technical Journal* 5.3: 8-20.

Hakes, David T. 1972. "Effects of Reducing Complement Constructions on Sentence Comprehension." *Journal of Verbal Learning and Verbal Behavior* 11.3 (June): 278-286.

Hakes, David T., and Helen S. Cairns. 1970. "Sentence Comprehension and Relative Pronouns." *Perception & Psychophysics* 8.1: 5-8.

Hakes, David T., and Donald J. Foss. 1970. "Decision Processes during Sentence Comprehension: Effects of Surface Structure Reconsidered." *Perception & Psychophysics* 8.6 (December): 413-416.

Hakes, David T., Judith S. Evans, and Linda L. Brannon. 1976. "Understanding Sentences with Relative Clauses." *Memory and Cognition* 4.3 (December): 283-290.

Hargis, Gretchen, et al. 2004. *Developing Quality Technical Information: A Handbook for Writers and Editors*, 2nd ed. Upper Saddle River, NJ: Prentice Hall Professional Technical Reference.

Harkus, Susan. 2001. *Writing for Translation*. ForeignExchangeTranslations. Available at http://www.multilingualwebmaster.com/library/writing-TR.html.

Hartshorn, Roy W. 1987. "Designing Information for the World." *Proceedings of the 34th International Technical Communication Conference*, WE186-WE189. Washington, D.C.: Society for Technical Communication.

Hatch, E., P. Polin, and S. Part. 1974. "Acoustic Scanning and Syntactic Processing: Three Experiments—First and Second Language Learners." *Journal of Reading Behavior* 6.3 (September): 275-285.

Hayes, Phil, Steve Maxwell, and Linda Schmandt. 1996. "Controlled English Advantages for Translated and Original English Documents." *Proceedings of the First International Workshop on Controlled Language Applications* (CLAW96). Leuven, Belgium: Katholieke Universiteit Leuven Centre for Computational Linguistics, March 26-27, 1996, pp. 84-92.

Hinson, Don E. 1991. "Simplified English—Is It Really Simple?" *Proceedings of the 38th International Technical Communication Conference*, WE33-WE36. Washington: Society for Technical Communication.

Hodgkinson, Richard, and John Hughes. 1982. "Developing Wordless Instructions: A Case History." *IEEE Transactions on Professional Communication* 25.2 (June): 74-79.

Hofmann, Patrick. 1998. "Wordless Manuals: Replacing Words with Pictures." *Proceedings of the 45th Annual Society for Technical Communication Conference*, 421-422. Arlington, VA: Society for Technical Communication.

Hoft, Nancy L. 1995. *International Technical Communication: How to Export Information about High Technology*. New York: John Wiley & Sons.

Holden, Norm. 1980. "Translations for the Rest of the World." *Technical Communication* 27.4: 23-25.

Hornby, Albert Sydney, and Sally Wehmeier. 2005. *The Oxford Advanced Learner's Dictionary of Current English*. 7th edition. London: Oxford University Press.

Huggins, A.W.F., and Marilyn Jager Adams. 1980. "Syntactic Aspects of Reading Comprehension." In *Theoretical Issues in Reading Comprehension*, edited by Rand J. Spiro, Bertram C. Bruce, and William F. Brewer. Hillsdale, NJ: Lawrence Erlbaum Associates, 87-112.

Huijsen, Willem-Olaf. 1998. "Controlled Language–An Introduction." *Proceedings of the Second International Workshop on Controlled Language Applications* (CLAW 98), May 21-22, 1998, Pittsburgh, PA., pp. 1-15.

Hunt, Peter, and John Kirkman. 1986. "The Problems of Distorted English in Computer Documentation." *Technical Communication* 33.3 (3rd Quarter): 150-156.

Hutchins, W. John. 1992. "Why Computers Do Not Translate Better." *Translating and the Computer 13 Conference Proceedings*. London: Aslib. Available at http://www.hutchinsweb.me.uk/Aslib-1991.pdf.

Hutchins, W. John. 1999. "The development and use of machine translation and computer-based translation tools." Available at http://www.hutchinsweb.me.uk/Beijing-1999.pdf.

Jones, A.R. 1996. "Tips on Preparing Documents for Translation." *Global Talk* (newsletter of the STC International Technical Communication SIG) 4.2: 4+.

Just, Marcel Adam, and Patricia A. Carpenter. 1987. *The Psychology of Reading and Language Comprehension*. Boston: Allyn and Bacon.

Kimball, John. 1973. "Seven Principles of Surface Structure Parsing in Natural Language." *Cognition* 2.1: 15-47.

Kirkman, John, Christine Snow, and Ian Watson. 1978. "Controlled English as an Alternative to Multiple Translations." *IEEE Transactions on Professional Communication* 21.4: 159-161.

Klein, Fred. 1988. "Appendix A: How to Prepare Documentation for International Marketing." In *Translation in Technical Communication*, edited by Fred Klein, 108-119. Washington, D.C.: Society for Technical Communication.

Kleinman, Joseph M. 1982. "A Limited-word Technical Dictionary for Technical Manuals." *Technical Communication* 29.1 (1st Quarter): 16-19.

Koch, Benjamin C. "Cost Control for Online Help Localization." *Intercom* 50.5 (May): 12-14.

Kohl, John R. 1990. "Article Usage." The Writing Center at Rensselaer Polytechnic Institute. Available at http://www.ccp.rpi.edu/esl.html.

Kohl, John R. 1999. "Improving Translatability and Readability with Syntactic Cues." *Technical Communication* 46.2 (May): 149-166.

Kohl, John R. 2007. "Assisted Writing and Editing at SAS." *ClientSideNews Magazine* 7.8 (August): 7-10. Available at http://www.clientsidenews.com/downloads/CSNV7I8.pdf.

Kulik, Ann B. 1995. "How the Tech Writer Improves Translation Results." *Global Talk* (newsletter of the STC International Technical Communication SIG) 3.1: 9-10.

Leech, Geoffrey, 2001. Paul Rayson, and Andrew Wilson. *Word Frequencies in Written and Spoken English: Based on the British National Corpus*. Harlow: Longman, Companion Web site available at http://ucrel.lancs.ac.uk/bncfreq/.

Locke, Nancy A. 2003. "Graphic Design with the World in Mind." *Intercom* 50.5 (May): 4-7.

Maaks, Betsy M. 2003. "Translation Stumbling Blocks." *Intercom* 50.5 (May): 8-9.

Macnamara, John. 1970. "Comparative Studies of Reading and Problem Solving in Two Languages." *TESOL Quarterly* 4.2: 107-116.

Martin, Edwin, and Kelyn H. Roberts. 1966. "Grammatical Factors in Sentence Retention." *Journal of Verbal Learning and Verbal Behavior* 5: 211-218.

Massion, François. 2007 "Terminology management: a luxury or a necessity?" *MultiLingual* 18.3 (April/May): 47-50.

Matkowski, Betty, and Gretchen Coar. 1983. "Translation: Telling the World about Your Products." *Proceedings of the 30th International Technical Communication Conference*, WE97-W100. Washington, D.C.: Society for Technical Communication.

Megathlin, Barclay A., and Robin S. Langford. 1991. "Controlling the Unruly: Terminology." *Proceedings of the 38th International Technical Communication Conference*, WE22-WE24. Washington, D.C.: Society for Technical Communication.

Merriam-Webster, Incorporated. 2005. *Merriam-Webster Online Dictionary*. http://www.merriam-webster.com.

Meyer, Charles F. 1987. *A Linguistic Study of American Punctuation*. New York: Peter Lang.

Microsoft Corporation. 2004. *Microsoft Manual of Style for Technical Publications*, 3rd ed. Redmond, WA: Microsoft Press.

Murphy, Dawn, Jane Mason, and Stuart Sklair. 1998. "Improving Translation at the Source." *Translating and the Computer 20 Conference Proceedings*. London: Aslib.

Nadziejka, David E. 1991. "Term Talk." *Technical Communication* 38.3 (3rd Quarter): 394-395.

Nielan, Cate. 2000. "Arthur Levitt and the SEC: Promoting Plain English." *Intercom* 47.9 (November): 17-18.

O'Brien, Sharon. 2003. "Controlling Controlled English: An Analysis of Several Controlled Language Rule Sets." *Proceedings of EAMT-CLAW03, Controlled Language Translation*, May 15-17, Dublin City University, 105-114.

O'Brien, Sharon, and Johann Roturier. 2007. "How Portable Are Controlled Language Rules? A Comparison of Two Empirical MT Studies." In Maegaard, Bente, ed., *Machine Translation Summit XI*. September 10-14, 2007, Copenhagen, 345-352. Available at http://www.mt-archive.info/MTS-2007-OBrien.pdf.

Ogden, C.K. 1932. *Basic English: A General Introduction with Rules and Grammar*. 3rd ed. London: K. Paul, Trench, Trubner & Co., Ltd.

Ogden, C.K. 1942. *The General Basic English Dictionary, giving more than 40,000 senses of over 20,000 words, in Basic English*. New York: W.W. Norton.

Olsen, Mary. 1984. "Terminology in the Computer Industry: Wading through the Slough of Despair." *Proceedings of the 31st International Technical Communication Conference*, WE200-WE201. Washington, D.C.: Society for Technical Communication.

Papineni, Kishore, Salim Roukos, Todd Ward, Wei-Jing Zhu. 2002. "BLEU: A Method for Automatic Evaluation of Machine Translation." In *Proceedings of the 40th Annual Meeting of the Association for Computational Linguistics*, Philadelphia, PA, July 2002, pp. 311-318.

Pearsall, Charles Robert. 1988. "John Deere Minimum Word Operator's Manual." In *Translation in Technical Communication*, edited by Fred Klein, 11-16. Washington: Society for Technical Communication.

Perfetti, Charles A., Sylvia Beverly, Laura Bell, Kimberly Rodgers, and Robert Faux. 1987. "Comprehending Newspaper Headlines." *Journal of Memory and Language* 26.6 (December): 692-713.

Peterson, D.A.T. 1990. "Developing a Simplified English Vocabulary." *Technical Communication* 37.2 (2nd Quarter): 130-133.

Phillips, WandaJane. 1991. "Controlling Terminology for Translation." *Proceedings of the 38th International Technical Communication Conference*, WE19-WE21. Washington, D.C.: Society for Technical Communication.

Plain Language Action and Information Network. *Federal Plain Language Guidelines.* Available at http://www.plainlanguage.gov/howto/guidelines/bigdoc/TOC.cfm.

Platzach, Christer. 1974. *Spraket och lasbarheten* (with English summary). Lund, Sweden: Gleerup.

Potsus, Whitney Beth, and Kaarina Kvaavik. 2001. "Is Your Documentation Translation-Ready?" *Intercom* 48.5 (May): 12-17.

Quirk, Randolph, et al. 1972. *A Grammar of Contemporary English.* Harlow: Longman.

Quirk, Randolph, and Sidney Greenbaum. 1973. *A Concise Grammar of Contemporary English.* New York: Harcourt Brace Jovanovich.

Rayner, K., M. Carlson, and L. Frazier. 1983. "The Interaction of Syntax and Semantics during Sentence Processing: Eye Movements in the Analysis of Semantically Biased Sentences." *Journal of Verbal Learning and Verbal Behavior* 22: 358-374.

Reuther, Ursula. 2003. "Two in One–Can It Work? Readability and Translatability by Means of Controlled Language." *Proceedings of EAMT-CLAW03, Controlled Language Translation*, 15-17 May 2003, Dublin City University, pp.124-132.

Roturier, Johann. 2004. "Assessing a Set of Controlled Language Rules: Can They Improve the Performance of Commercial Machine Translation Systems?" *Proceedings of Translating and the Computer 26*. London: Aslib.

Rushanan, Valerie. 2007. "Editing for International Audiences." *Intercom* 54.10 (December): 16-19.

Rychtyckyj, Nestor. 2002. "An Assessment of Machine Translation for Vehicle Assembly Process Planning at Ford Motor Company." *Proceedings of the 5th Conference of the Association for Machine Translation in the Americas*, AMTA 2002, Tiburon, CA., pp. 207-215. Available at http://www.cs.wayne.edu/~ner/amta-2002.pdf.

Sakiey, E., and E. Fry. 1979. *3000 Instant Words*. Highland Park, NJ: Dreier Educational Systems.

Sanderlin, Stacey. 1988. "Preparing Instruction Manuals for Non-English Readers." *Technical Communication* 35.2 (2nd Quarter): 96-100.

Schachtl, Stefanie. 1996. "Requirements for Controlled German in Industrial Applications." *Proceedings of the First International Workshop on Controlled Language Applications* (CLAW '96), March 26-27, 1996, Leuven, Belgium, pp. 143-149.

Schluroff, Michael, Thomas E. Zimmermann, R. B. Freeman Jr., Klaus Hofmeister, Thomas Lorscheid, and Arno Weber. 1986. "Pupillary Responses to Syntactic Ambiguity of Sentences." *Brain and Language* 27.2 (March): 322-344.

Sheridan, E.F. 2001. "Cross-cultural Web Site Design." *MultiLingual Computing & Technology* 12.7 (October/November): 53-56.

Smith, Frank. 1988. *Understanding Reading: A Psycholinguistic Analysis of Reading and Learning to Read*. 4th ed. Hillsdale, NJ: Erlbaum, p. 150.

Spaggiari, Laurent, Florence Beaujard, and Emmanuelle Cannesson. 2003. "A Controlled Language at Airbus." *Proceedings of EAMT-CLAW03, Controlled Language Translation*, May 15-17, Dublin City University, pp. 151-159.

St. Amant, Kirk R. 2003. "Designing Web Sites for International Audiences." *Intercom* 50.5 (May): 15-18.

Strong, Kathy L. 1983. "Kodak International Service Language." *Technical Communication* 30.2: 20-22.

Sullivan, Dan. 2001. "Machine Translation: Is It Good Enough?" *e-Business Advisor*. June 2001: 32, 36-37.

Swan, Michael, Bernard Smith, and Penny Ur. 2001. *Learner English: A Teacher's Guide to Interference and Other Problems*. 2nd Edition. Cambridge, UK: Cambridge University Press.

Swenson, Lynne V. 1988. "How to Make (American) English Documents Easy to Translate." In *Translation in Technical Communication*, edited by Fred Klein, 93-95. Washington, D.C.: Society for Technical Communication.

Thomas, Margaret, Gloria Jaffe, J. Peter Kincaid, and Yvette Stees. 1992. "Learning to Use Simplified English: A Preliminary Study." *Technical Communication* 39.1 0 (1st Quarter): 69-73.

Twyman, M. 1985. "Using Pictorial Language: A Discussion of the Dimensions of the Problem." In *Designing Usable Texts*, edited by Thomas M. Duffy and Robert Walker. Orlando: Academic Press.

Underwood, Nancy, and Bart Jongejan. 2001. "Translatability Checker: A Tool to Help Decide Whether to Use MT." *Proceedings of MT Summit VIII*, Santiago de Compostela, Galicia, Spain, 18-22 Sep 2001, pp. 363-368.

Unwalla, Mike. 2004. "AECMA Simplified English." *Communicator*. Winter 2004: 34-35.

Velte, Charles E. 1990. "Is Bad Breadth Preventing Effective Translation of Your Text?" *Proceedings of the 37th International Technical Communication Conference*, RT5-RT7. Washington, D.C.: Society for Technical Communication.

Vogt, Herbert E. 1986. "Graphic Ways to Eliminate Problems Associated with Translating Technical Documentation." *Proceedings of the 33rd International Technical Communication Conference*, 330-333. Washington, D.C.: Society for Technical Communication.

Warburton, Kara. 2003. "The Terms of Business: Saving Money through Terminology Management." *Globalization Insider* Issue 4.3 (November). Available at http://www.lisa.org/globalizationinsider/2003/11/the_terms_of_bu.html.

Weiss, Edmond H. 1998. "Twenty-Five Tactics to 'Internationalize' Your English." *Intercom* 45.5: 11-15.

White, E. N. 1981. "Using Controlled Languages for Effective Communication." *Proceedings of the 28th International Technical Communication Conference*, E110-E113. Washington, D.C.: Society for Technical Communication.

Williams, Joseph M. 2005. *Style: Ten Lessons in Clarity and Grace*, 8th ed. New York: Pearson Longman.

Wittner, Janaina. 2007. "Unexpected ROI from terminology." *MultiLingual* 18.3 (April/May): 51-54.

Wojcik, Richard H., and Heather Holmback. 1996. "Getting a Controlled Language Off the Ground at Boeing." *Proceedings of the First International Workshop on Controlled Language Applications* (CLAW96). Leuven, Belgium: Katholieke Universiteit Leuven Centre for Computational Linguistics, March 26-27, 1996, pp. 22-31.

Wright, Sue Ellen, and Budin, Gerhard (eds.). 1997-2001. *Handbook of Terminology Management*, vols 1 and 2. John Benjamins: Amsterdam and Philadelphia.

Yunker, John. 2003. *Beyond Borders: Web Globalization Strategies*. Indianapolis: New Riders Press.

Zhang, Yu. 2006. "The Chinese Challenge: A Localization Project Manager's Guide." *The Globalization Insider* Issue 6 (August). Available at http://www.lisa.org/globalizationinsider/2006/08/the_chinese_cha.html.

Zinsser, William. 1998. *On Writing Well: The Classic Guide to Writing Nonfiction*, 6th ed, rev. and updated. New York: HarperCollins.

Index

A

B

C

E

Q

R

U

V

W

X

Comments or Questions?

If you have comments or questions about this book, you may contact the author through SAS as follows.

Mail: SAS Institute Inc.
SAS Press
Attn: <Author's name>
SAS Campus Drive
Cary, NC 27513

E-mail: saspress@sas.com

Fax: (919) 677-4444

Please include the title of the book in your correspondence.

CPSIA information can be obtained at www.ICGtesting.com
Printed in the USA
BVOW061838120712

295048BV00004B/3/P